On the Improvement of Combustion Engines with Waste Heat Recovery Systems in Mobile Applications

Zur Erlangung des akademischen Grades
Doktor der Ingenieurwissenschaften

der Fakultät für Maschinenbau
des Karlsruher Instituts für Technologie (KIT)

genehmigte
Dissertation

von

Dipl.-Ing. Thomas Matoušek
aus Erlangen

Tag der mündlichen Prüfung:	10.05.2019
Hauptreferent:	Prof. Dr. sc. techn. Thomas Koch
Korreferent:	Prof. Dr.-Ing. Peter Stephan

**Forschungsberichte aus dem
Institut für Kolbenmaschinen
Karlsruher Institut für Technologie (KIT)
Hrsg.: Prof. Dr. sc. techn. Thomas Koch**

Bibliografische Information der Deutschen Nationalbibliothek

Die Deutsche Nationalbibliothek verzeichnet diese Publikation in der
Deutschen Nationalbibliografie; detaillierte bibliografische Daten sind
im Internet über http://dnb.d-nb.de abrufbar.

ISBN 978-3-8325-4956-5
ISSN 1615-2980

Logos Verlag Berlin GmbH
Comeniushof, Gubener Str. 47,
10243 Berlin
Tel.: +49 030 42 85 10 90
Fax: +49 030 42 85 10 92
INTERNET: http://www.logos-verlag.de

Vorwort des Herausgebers

Die Komplexität des verbrennungsmotorischen Antriebes ist seit über 100 Jahren Antrieb für kontinuierliche Aktivitäten im Bereich der Grundlagenforschung sowie der anwendungsorientierten Entwicklung. Die Kombination eines instationären, thermodynamischen Prozesses mit einem chemisch reaktiven und hochturbulenten Gemisch, welches in intensiver Wechselwirkung mit einer Mehrphasenströmung steht, stellt den technologisch anspruchsvollsten Anwendungsfall dar. Gleichzeitig ist das Produkt des Verbrennungsmotors aufgrund seiner vielseitigen Einsetzbarkeit und zahlreicher Produktvorteile für sehr viele Anwendungen annähernd konkurrenzlos. Nun steht der Verbrennungsmotor insbesondere aufgrund der Abgasemissionen im Blickpunkt des öffentlichen Interesses. Vor diesem Hintergrund ist eine weitere und kontinuierliche Verbesserung der Produkteigenschaften des Verbrennungsmotors unabdingbar.

Am Institut für Kolbenmaschinen am Karlsruher Institut für Technologie wird deshalb intensiv an der Weiterentwicklung des Verbrennungsmotors geforscht. Übergeordnetes Ziel dieser Forschungsaktivitäten ist die Konzentration auf drei Entwicklungsschwerpunkte. Zum einen ist die weitere Reduzierung der Emissionen des Verbrennungsmotors, die bereits im Verlauf der letzten beiden Dekaden um circa zwei Größenordnungen reduziert werden konnten, aufzuführen. Zum zweiten ist die langfristige Umstellung der Kraftstoffe auf eine nachhaltige Basis Ziel der verbrennungsmotorischen Forschungsaktivitäten. Diese Aktivitäten fokussieren gleichzeitig auf eine weitere Wirkungsgradsteigerung des Verbrennungsmotors. Der dritte Entwicklungsschwerpunkt zielt auf eine Systemverbesserung. Motivation ist beispielsweise eine Kostenreduzierung, Systemvereinfachung oder Robustheitssteigerung von technischen Lösungen. Bei den meisten Fragestellungen wird aus dem Dreiklang aus Grundlagenexperiment, Prüfstandversuch und Simulation eine technische Lösung erarbeitet.

Die Arbeit an diesen Entwicklungsschwerpunkten bestimmt die Forschungs- und Entwicklungsaktivitäten des Instituts. Hierbei ist eine gesunde Mischung aus grundlagenorientierter Forschung und anwendungsorientierter Entwicklungsarbeit der Schlüssel für ein erfolgreiches Wirken. In nationalen als auch internationalen Vorhaben sind wir bestrebt, einen wissenschaftlich wertvollen Beitrag zur erfolgreichen Weiterentwicklung des Verbrennungsmotors beizusteuern. Sowohl Industriekooperationen als auch öffentlich geförderte Forschungsaktivitäten sind hierbei die Grundlage guter universitärer Forschung.

Zur Diskussion der erarbeiteten Ergebnisse und Erkenntnisse dient diese Schriftenreihe, in der die Dissertationen des Instituts für Kolbenmaschinen verfasst sind. In dieser Sammlung sind somit die wesentlichen Ausarbeitungen des Instituts niedergeschrieben. Natürlich werden darüber hinaus auch Publikationen auf Konferenzen und in Fachzeitschriften veröffentlicht. Präsenz in der Fachwelt erarbeiten wir uns zudem durch die Einreichung von Erfindungsmeldungen und dem damit verknüpften Streben nach Patenten. Diese Aktivitäten sind jedoch erst das Resultat von vorgelagerter und erfolgreicher Grundlagenforschung.

Jeder Doktorand am Institut beschäftigt sich mit Fragestellungen von ausgeprägter gesellschaftlicher Relevanz. Insbesondere Nachhaltigkeit und Umweltschutz als Triebfedern des ingenieurwissenschaftlichen Handelns sind die Motivation unserer Aktivität. Gleichzeitig kann er nach

Beendigung seiner Promotion mit einer sehr guten Ausbildung in der Industrie oder Forschungslandschaft wichtige Beiträge leisten.

Im vorliegenden Band berichtet Herr Matousek über experimentelle und simulative Untersuchungen zum Thema Restwärmenutzung verbrennungsmotorischer Abgase in der PKW-Anwendung. Es wurde speziell für diese Aufgabe ein Motorenprüfstand aufgebaut der mit einem seriennahen Restwärmenutzungssystem ausgestattet wurde. Als Arbeitsprinzip für das Restwärmenutzungssystem wurde der thermodynamische Kreisprozess nach Rankine gewählt: ein Arbeitsmedium wird von einer Pumpe zu einem Verdampfer gefördert, der entstandene Dampf treibt eine Expansionsmaschine an und wird anschließend wieder kondensiert. Um die Ergebnisse auf weitere Restwärmenutzungssysteme zu übertragen wurde ein Simulationsmodell aufgebaut, das Verbrennungsmotor und Rankine-System abbildet.

Aufgrund der verbrennungsmotorischen Randbedingungen und den Einschränkungen eines Kraftfahrzeugs ergeben sich neue Betriebsbedingungen für Rankine-Systeme. Diese hat Herr Matousek im Rahmen seines Forschungsprojektes untersucht. Thematisch hat er sich dabei zuerst mit den Betriebsparametern im stationären Zustand des Systems befasst.

Die Optimierung der Systemparameter bietet das Potential den Wirkungsgrad eines laufenden Systems zu verbessern, in dem jeweils der aktuellen Situation angepasste optimale Parameter eingestellt werden. Neben dem grundlegenden Aufbau nach Rankine hat Herr Matousek auch andere Kreislaufvarianten untersucht, die einen weiteren Wärmeeintrag in System bieten und somit die Leistungsabgabe steigern können. Als letzten operativen Einfluss auf ein Restwärmenutzungssystem wurde die Bedeutung der motorischen Betriebsparameter analysiert. Hierbei ging es einerseits um die Schaffung von Grundlagen, aber auch um die Suche nach möglichen Potentialen. Diese könnten in Änderungen des motorischen Betriebs gefunden werden, die geringere Nachteile für den Verbrennungsmotor haben, als sie Vorteile für das Restwärmenutzungssystem bieten. Somit ließe sich der Gesamtsystemwirkungsgrad steigern.

Die im stationären Betrieb gewonnen Erkenntnisse wurden auf den dynamischen Betrieb übertragen. Als wichtige Variante dessen, wurde der Kaltstart gewählt. Dieser konnte dank der Tieftemperaturfähigkeit des Prüfstands bei Temperaturen von bis zu -10°C dargestellt werden. Herr Matousek hat hier Grundlagen zum Kaltstartverhalten eines Rankine-Systems geschaffen und die Einflüsse von Systemparametern und Kreislaufvarianten dargestellt. Als abschließende Facette der Gesamtsystembetrachtung und der gewonnen Erkenntnisse wurden mögliche Synergien durch verschiedene Arten der Kühlmittelkreislaufanbindung des Kondensators untersucht. Die verbleibende Wärme die dem Abgas entnommen wurde fällt am Kondensator an und bietet die Möglichkeit den Kaltstart des Fahrzeugs zu verbessern.

Somit sind in der vorliegenden Arbeit übertragbare Ergebnisse zum gemeinsamen Betrieb von Restwärmenutzungssystem und Verbrennungsmotoren geschaffen worden, die helfen können, den Gesamtsystemwirkungsgrad von Kraftfahrzeugen zu steigern.

Karlsruhe, im Mai 2019   Prof. Dr. sc. techn. Thomas Koch

Vorwort des Autors

Die vorliegende Arbeit entstand während meiner Tätigkeit als wissenschaftlicher Mitarbeiter am Institut für Kolbenmaschinen des Karlsruher Instituts für Technologie. Sie ging aus einem von 2014 bis 2017 geförderten Projekt der Friedrich-und-Elisabeth Boysen Stiftung für Forschung und Innovation zum Thema "Analyse und Optimierung des Thermomanagements eines Verbrennungsmotors zur Optimierung der Leistungsabgabe von Rankine-Kreisläufen zur Restwärmenutzung" hervor.

Mein besonderer Dank gilt an erster Stelle Herrn Prof. Dr. sc. techn. Thomas Koch für das mir entgegengebrachte Vertrauen und die mir gebotene Freiheit bei der Gestaltung und Durchführung der Arbeit. Für die Übernahme des Korreferats und das große Interesse an meiner Arbeit danke ich Herrn Prof. Dr.-Ing. Peter Stephan. Prof. Dr.-Ing. Ulrich Spicher möchte ich danken, dass er mir den Weg an das Institut ermöglicht hat.
Nur durch die Unterstützung mit Material und Knowhow von Jan Gärtner war es überhaupt möglich, den Prüfstand für meine Arbeit aufzubauen und zu betreiben - dafür möchte ich mich ausdrücklich bedanken. Im Rahmen des Aufbaus eben dieses Prüfstandes haben Dominik Mall und Andreas Hartmann mich maßgeblich beraten. Für diesen Support und für den Wissenstransfer bedanke ich mich. Ein steter Unterstützer war mein Projektpartner Michael Bens, der von mir nicht nur frohe Botschaften erhielt, aber mit dem es immer eine Freude war sich thematisch und nicht-thematisch auszutauschen.
Die Ergebnisse meiner Forschung durfte ich auf einer Vielzahl an nationalen und internationalen Tagungen vorstellen, dabei war mein häufigster Begleiter Dr.-Ing. Amin Velji. An die gemeinsamen Reisen, Gespräche und Erlebnisse erinnere ich mich immer gerne. Dr.-Ing. Heiko Kubach, Dr.-Ing. Kai W. Beck und Dr.-Ing. Uwe Wagner waren meine Gruppenleiter, in dieser Funktion haben sie mich nie im Stich gelassen.
Ich danke allen Kollegen, die zum Gelingen dieser Arbeit, sowie zu meiner Zeit am Institut beigetragen haben. Natürlich zählen dazu auch die Aktivitäten außerhalb des Institutsalltages: Motorradtouren, Fußballspiele, Kap- und Vogelbesuche und viele mehr. Diese waren willkommene Ablenkungen vom Arbeitsstress, an die ich gerne zurückdenke. Einigen Kollegen möchte ich dabei besonders danken. Gemeinsam mit mir zusammen durfte Frank Martin Stahl viele Projekte und Aufgaben im Bereich Restwärmenutzung bearbeiten. Wir haben zusammen viel gelacht, gelernt und geschuftet, aber es war mir eine Freude mit ihm zu arbeiten. Besonders bedanken möchte ich mich auch bei meinem langjährigen Bürokollegen Marius Neurohr. Er war immer ein Quell des Wissens was verbrennungsmotorische Themen und Schabernack anging. Michael Rößler und Panagiotis Maniatis waren meine Lieblings-Nicht-Büro-Kollegen. Beide sind Experten auf dem Gebiet Verbrennungsmotoren. Michael schwankt zwischen Genie und Wahnsinn und Pana stellt die besten Fragen. Meinem französischen Kollegen Christoph Pfister danke ich, dass ich nun sagen kann, dass ich weiß wie Pferdefleisch und Schnecken schmecken. Meinen Kollegen aus MOT Zeiten, Ina Volz, Dr.-Ing. Clemens Hampe und Fabian Titus möchte ich für die gemeinsamen Ausflüge und die verbrachten Stunden im MOT-Dachboden danken. Dr.-Ing. Daniel Ghebru war mein Diplomarbeitsbetreuer und hat mich ans Institut geholt. Er war mir stets ein Vorbild was Forscherdrang und Arbeitsmoral angeht. Dr.-Ing. Helge Dageförde hat mir beigebracht, wie man Forschungsanträge schreibt und was es bedeutet Partikelemissi-

onen zu messen. Ivica Kraljevic danke ich für den inter-institutionellen Austausch zum Thema Restwärmenutzung, der mir oft einen Blick über den Tellerrand erlaubt hat. Meinem Freundeskreis aus Erlangen/Nürnberg danke ich für die regelmäßigen Besuche und Freizeitaktivitäten im fernen Karlsruhe.

Im technischen Bereich möchte ich einigen Kollegen meinen Dank aussprechen, die beim Gelingen dieser Arbeit geholfen haben. Helge Rosenthal war bei sämtlichen technischen Fragestellungen eine hilfreiche und hilfsbereite Anlaufstelle. Gregor Rosbach hat so gut wie jede Schweißnaht am Prüfstand durchgeführt. Ihm ist der Grundaufbau des Motors zu verdanken. Was sonst getan werden musste, hat Christian Stahl entweder schnell durchführen lassen oder zuverlässig selbst erledigt. Für die schnelle und präzise Fertigung zahlreicher kleiner und großer Versuchsteile danke ich der gesamten Fertigungsabteilung um Ernst Hummel. Herr Joachim Strauch hat mich mit seinem Fachwissen begeistert und hat die Automatisierung des Prüfstands übernommen. Walter Trepka hatte immer einen schlechten Witz auf Lager und konnte bei elektrischen Problemen helfen.

Im Laufe meiner Jahre am Institut durfte ich mit einer Vielzahl an Studenten zusammenarbeiten, deren Auflistung den Rahmen dieses Vorwortes übersteigt. Alleine zu Zeiten meines Dauerlaufprojektes dürften es um die 30 gewesen sein. Ein paar Studenten, die über längere Zeit wesentlich zum Gelingen dieser Arbeit beigetragen haben, möchte ich dennoch gesondert danken. Paul Lagaly hat viele Stunden mit dem Auf- und Abbau meiner Prüfstände verbracht, ebenso wie mit der Verbesserung und Erweiterung vieler meiner Simulationsmodelle. Zum Dank für seine geordnete Arbeitsweise ist er mittlerweile auch ein Kollege am Institut. Andreas Röder hat sehr eigenständig einen außerordentlichen Aufbau eines Fahrwindgenerators vollbracht. Corentin Charles hat mir während meiner Zeit mit gebrochener rechter Hand eben diese ersetzt - selbst zu unfranzösischen Uhrzeiten und Tagen. Janina Mattes und Matteo Vogel haben für mich die Grundlagen der GT-Modelle geschaffen in Zeiten als ich an allen Fronten kämpfen musste. Tobias Schneider hat den Grundaufbau des Motors und Kreislaufs mitbetreut und war immer interessiert. Patrick Wunsch und Jana Aberham haben mir lange Zeit als Hiwis gute Dienste geleistet und durften mein Projekt in unterschiedlichsten Phasen begleiten und haben immer tatkräftig geholfen.

Sehr großer Dank gilt meiner Mutter Hannelore Meisel, die für mich immer da war, wenn ich sie gebraucht habe. Sie war mir stets ein sicherer Hafen und wusste in technischen Dingen immer Rat. Eine bessere Mutter kann sich kein Sohn wünschen. Meinem Vater möchte ich danken, dass er mir immer den Wert guter Bildung vermittelt hat. Er hat mich immer zu mehr angetrieben. Leider konnte er die Fertigstellung dieser Arbeit nicht mehr miterleben. Meinen Geschwistern Michaela, Petra und Robert möchte ich dafür danken, dass ich eine so schöne Kindheit erleben durfte und dass ich dadurch zu dem Mensch wurde, der ich nun bin. Abschließend möchte ich mich bei meiner Frau Anne Matoušek bedanken. Sie ist die größte Unterstützung für diese Arbeit gewesen. Sowohl moralisch, inhaltlich als auch manuell hat sie, selbst während ihrer Schwangerschaft einen Betrag geleistet, der nicht aufzuwiegen ist. Nur durch ihren Beistand war es mir möglich die nötigen Freiräume zu haben, um dieses Werk zu vollenden.

Karlsruhe, im Mai 2019   Thomas Matoušek

Contents

Nomenclature

Physical Quantities

Symbol	Unit	Description
α	°CA	Crank angle
η	%	Efficiency
Π	-	Pressure ratio
λ	W/(m·K)	Thermal conductivity
λ	-	Air-fuel equivalence ratio
A	m²	Area
c_p	kJ/(kg·K)	Specific heat capacity
F	N	Force
H_U	MJ/kg	Lower heating value
H, h	J,J/kg	Enthalpy, specific enthalpy
m	kg/h	Mass
M	Nm	Torque
M_d	Nm	Engine torque
n	rpm	Speed
p	bar	Pressure
P	kW	Power
P_e	kW	Effective engine power
Q, q	J,J/kg	Heat, specific heat
r	m	Distance
S, s	J/K,J/(kg·K)	Entropy, specific entropy
T	°C	Temperature
t	s	Time
U, u	J,J/kg	Internal energy, specific internal energy
v	km/h	Velocity
$\dot{V}$	lpm	Volume flow
V	l,m³	Volume
W, w	J, J/kg	Work, specific work
x	-	Steam quality

Abbreviations and Indices

Abbreviation	Description
1	After condenser
2	After pump
3	After exhaust gas heat exchanger
4	After expander
a	After
AC	Air conditioning
act	Actual value

Abbreviation	**Description**
Amb	Ambient
b	Before
BTDC	Before top dead center
Cab	Cabin
CAC	Charge air cooler
CAE	Computer-aided engineering
CO	Carbon monoxide
Cond	Condenser
DC	Direct current
Dev	Deviation
Des	Desired
ECU	Electronic Control Unit
EG	Exhaust gas
EGR	Exhaust gas recirculation
EGX	Exhaust gas heat exchanger
el	Electric
Env	Environment
EPA	United States Environmental Protection Agency
ETK	Emulatortastkopf
Exp	Expander
Fl	Fluid
GPF	Gasoline particulate filter
HC	Hydrocarbons
HX	Heat exchanger
in	In
IT	Ignition timing
low	Low-pressure part of the system
man	Manual
NEDC	New European driving cycle
NO_x	Nitrogen oxides
OP	Operating point
opt	Optimal
out	Out
pf	Partial flow
PID	Proportional–integral–derivative
PM, PN	Particulate matter, particulate number
Preh	Preheater
rev	Reversible
Set	Setpoint
SOI	Start of injection
SI	Système international (d'unités)
TC	Turbo-compound
TDC	Top dead center
TEG	Thermoelectric generator
TACG	Thermoacoustic generator
Theo	Theoretically
Turbo	Turbocharger
VDA	Verband der Automobilindustrie e. V.
W	Wall
WF	Working fluid
WHR	Waste heat recovery
WLTC	Worldwide harmonized Light vehicles Test Cycle

1 Introduction

1.1 Background

CO_2 was recognized as one of the main contributors to global warming at the 21st Climate Change Conference in Paris. It was agreed that the emission of greenhouse gases such as CO_2 has to be reduced in order to limit global warming to 2 K [25]. Since the beginning of industrialization the emission rate of CO_2 is steadily increasing [26]. To reverse this trend new technologies have to be developed that reduce the CO_2 production.

Traffic emissions (incl. passenger and commercial vehicles) are responsible for 17 % of the world wide CO_2 emissions [10]. Most of them are equipped with combustion engines which are and will be the main drive for individual mobility. Despite efforts to vitalize public transportation, in 2014 76 % of passenger traffic was still individual (motorized) [106]. As long as people still use their freedom of individual mobility, not even a technological switch to electric cars, which is so often propagated, would reduce CO_2 emissions [97]. Yet combustion engines will still be the most efficient form of propulsion when evaluating the system "car" holistically. Their ability to efficiently deliver mechanical power and to provide heat for the passengers makes them superior to any other kind of propulsion. As they will be used in the future as well, they have to be improved to emit less CO_2 and to keep the climate goal reachable.

The EU is trying to force improvement in combustion engines through stricter legislation, which is an effective way as the advancements through the Euro 1-5 emission standards prove. The latest legislation sets emission limits an punishes the production of CO_2 above these (See table 1.1).

Table 1.1: Limits and penalties for exceeding the limits of CO_2-emissions in the EU [25]

Limit 2015:	130 g/km	Limit 2020:	95 g/km

Until 2018		Starting 2018	
1 g	5 €/(car · g/km)	> 1 g	95 €/(car · g/km)
2 g	15 €/(car · g/km)		
3 g	25 €/(car · g/km)		
> 4 g	95 €/(car · g/km)		

With a 2020 emission limit of 95 g/km and a penalty of 95 €/(car · g/km) the cost of exceeding the limit can easily reach hundreds of Euros per car. This means that car manufactures should be able to invest more money into fuel-saving technologies without losing any competitive advantage, as competitors that don't invest in improvement have to pay the fine. Or paraphrased in other words: new measures for increasing efficiency are becoming economical. [19]

1.2 Motivation

Most car manufactures focus their effort on increasing the efficiency of combustion engines, lightweight design, or on aerodynamics, but there are other technologies to lower the fuel consumption of the whole system "car". Research on the first three mentioned technologies is performed since the beginnings of the automobile and while there is still potential most of it is harvested [79]. The technology that this work is focused on, is a new approach that is not yet implemented in any series production car. This promising technology is the use of waste heat in the form of exhaust gas.

The exhaust gas contains about one third of the energy that is released by combusting the fuel and most of it is wasted [96]. Waste heat recovery (WHR) systems are employed to regain parts of that energy. The most promising WHR-technology is considered to be the steam cycle according to the design of Rankine [55], [102]. WHR-systems like Rankine-systems work based on the vaporization of a working fluid through the heat of the exhaust gas. The produced steam is used to drive an expansion machine which generates mechanical power. This power can be used to support the combustion engine and reduce its fuel consumption either directly or through the use of a generator by supplying electric energy.

The remaining steam has to be condensed in order to be recirculated. This process takes place in the condenser. The enthalpy of condensation is transferred to a heat sink. In a car this heat sink can either be the surrounding air or the cooling system of the car. This enthalpy could potentially be used to improve the cold start behavior of the engine and the passenger cabin.

The basic principle was found by William Rankine in the years 1850-1860 and has since been used to describe and improve the processes of steam driven power plants and engines. The research in the field of thermal recuperation in cars through waste heat recovery systems like Rankine-systems has gained new impulses in the last years. Research focuses mainly on commercial vehicles, because of fuel costs being the most significant contribution to the overall cost of operation in this application and any reduction of it can directly be translated into monetary savings for the owner. The development of Rankine-systems for commercial vehicles is quite advanced and production ready prototypes exist for most components [38]. As soon as fuel costs and system costs level out, these systems will find their way into production [77], [48].

The technology of WHR-systems for passenger cars is not as advanced as for commercial vehicles. Car manufactures have constructed and tested different system designs, built up from prototypes or from parts that come from other applications. These early studies showed promising results, but were not pursued up to series production, yet (see chapter 2.2.1). There are several reasons for this. The first reason is cost and the lack of pressure from legislation up to this point. The other reason is the need for further research.

The reason for the restrained research in this field can be attributed to many issues: the lack of knowledge of the ideal working fluid, uncertainty of the future of combustion engines and the remaining potential in improving the engine itself. Other improvements are seen to be easier and more cost-effective to deploy on a car at the moment [19]. But at some point all these technologies will be implemented and the cost of WHR-systems will come down to the reasonable area.

With the future inclusion of 48 V electrical systems, the integration of these systems will be much less intricate. The power for the pump can be supplied with smaller cables, the power output of the expander can be transformed to a more fitting level and the demand for electric power is generally higher. Electric coolant pumps, electric air conditioning (AC) compressors and electric turbochargers are future developments for hybrid cars that will benefit from the electric energy that can be supplied by WHR-systems.

Rudolf Diesel [16] stated that "no advantage is to be expected" by a "regenerator" that transforms the combustion waste heat for "propulsion purposes" and that the "minimal benefits" do not justify the complexity of one. Despite that, most car manufacturers see potential in the technology as the research of the last years proves. BMW developed a complex system that demonstrated major advantages on an engine test bench [29] and put a simpler version into a passenger car [28]. Audi also worked on a WHR-system based on the Rankine principle. They built a simpler system on a test bench using a gas burner as a heat source [57]. Daimler was able to create a working system that was included into a passenger car [42]. IAV created a Rankine-system with one heat exchanger ahead of the turbocharger and one after the catalyst, which helped reduce the need for fuel enrichment under full load. Their simulation results also indicated that synergies could be gained in combination with downsizing and the resulting increase in exhaust temperature [78]. Honda went with a similar approach and integrated the heat exchanger into the catalyst, this helped to increase the heat flow to their Rankine-system, however with an impact on the catalyst function. They also included dedicated channels into the cylinder head of the engine which acted as a preheater for the working fluid to improve the power output [22]. Only little details of this industrial research is published and most of the published work is concentrated on components and working fluid with mostly theoretical approaches or in form of simulations [8], [41], [87].

This work tries to present detailed information about the coherencies of a WHR-system and a combustion engine in a passenger car application on the example of a Rankine-system with a real experimental setup.

1.3 Objective & Structure

The objective of this work was to generate information about the performance and the influences of different system configurations and parameters for steady-state operation and cold start scenarios of a Rankine-system integrated in the exhaust system of an engine. Several tests were performed on a test bench that was specifically created for this purpose.

A modern 4-cylinder 2.0 l SI engine, which fulfills the EURO 6 standard, was chosen to conduct the tests on. A WHR-system consisting of components as close-to-production as possible was applied to the engine's exhaust system. The engine and the WHR-system were put on a test bench which provided stable boundary conditions and was capable of conditioning the chamber to temperatures between 20 to -10°C. A simulation model of the combustion engine and the Rankine-system was created to expand the experimental results.

The first tests were performed in order to find the basic influence of the main system operating parameters: mass flow, low-pressure and low-temperature. These investigations created the base for the following experiments and their interpretation. The system's operating parameters offer potential to increase the power output of a running system by adapting them to the current condition of the system. The most promising settings were chosen for the subsequent tests.

The potential use of additional heat sources was investigated on the basis of the previous results. Four different configurations were investigated to find the potential improvement in power output through further heat input and eventual improvements in efficiency. These included the use of a partial flow recuperator, a coolant aided preheater, a full flow recuperator and the use of the turbocharger waste heat.

The influence of engine operating parameters on the WHR-system was analyzed to generate basic understanding of their relevance and to find potential for improving the overall system. The potential was anticipated to be found by changing the combustion process without, or at least with only minor disadvantages for the engine's efficiency and overcompensating that by an increased power output of the WHR-system and therefore improving the overall efficiency of the car.

Another target was to display the warm-up behavior of WHR-systems and engines during cold start at different starting temperatures (-10° to +20°C) under realistic conditions for different coolant circuit configurations. Preliminary investigations were performed to identify the appropriate coolant volume flow rates and circuit configurations to improve the warm up of the desired component. The resulting temperatures of the components, the condenser heat flow and the power output of the system were the main focus points.

All previous observations were combined in the final investigations to try to display possible synergies between WHR-system and engine in cold start scenarios. By comparing different coolant circuit connections to the condenser and engine components, it was examined, whether the remaining heat, that can be gathered at the condenser, can be beneficial for the heat-up of the engine, its parts and/or the passenger cabin. This approach was chosen to depict the overall impact of a WHR-system in the transient operation of a cold start on fuel consumption and comfort.

2 Principles

This chapter serves to explain the basic principles that are relevant for understanding the subsequent chapters. Only principles that are needed or have to be defined will be discussed in detail.

2.1 Thermodynamic Principles

Thermodynamics is the study of the conversion of energy and the gain of usable work thereof, as it occurs in natural and technical processes [4]. The basic behavior of such conversions is explained by the first and second laws of thermodynamics. The science of thermodynamics allows the understanding and prediction of energy conversions by calculating and measuring the state functions of a system, like temperature, pressure, mass, volume, energy etc.. [4], [7]

2.1.1 State Quantities

The exact acquisition and assessment of a system demands knowledge of the state quantities of said system. These quantities describe the state of it and can be divided into outer quantities and inner quantities. Outer quantities (or mechanical quantities) describe the position and movement of a system. Inner quantities (or thermodynamic quantities) represent the properties of the content within the system borders: e.g. temperature, pressure or mass. [4], [65]
The two most important ones for this study will be discussed in the following.

Temperature is a base unit of the SI-system. It can be described as the average kinetic energy of the random motions of the electrons, atoms and molecules that make up matter. The faster their movement, the higher is the temperature. Temperature determines if a system is in thermal equilibrium or not. If a system has a higher temperature than another, it is denoted as hotter and the one with lower temperature is denoted colder. Heat can only be transferred if there is a temperature difference between systems. This happens until the heat has balanced the temperatures of both systems, at which point the systems are in thermal equilibrium. [65]

Pressure is a unit that is derived from other SI-units. Pressure is defined as the quotient of a force to the area that it is applied to. One can imagine that the motion and vibration of particles within a system transfer an impulse when getting in contact to the borders of the system. The microscopic stochastic distribution results macroscopically in an even force on that border. As temperature increases the kinetic energy of the particles, it also increases the pressure in a constant volume. Likewise the pressure is increased if the volume is reduced and the temperature is constant. [4], [65]

These quantities together with mass (m), volume (V) and the specific gas constant (R) can be brought into relation with the ideal gas law: [100]

$$p \cdot V = m \cdot R \cdot T \tag{2.1}$$

2.1.2 Internal Energy, Enthalpy and Entropy

Internal Energy
The amount of energy contained within the boundaries of a system is called the internal energy
of a system (U). It does not take any ambient conditions into account (kinetic energy, potential
energy, etc.) - only the energy within the system. For a closed system, this energy consists only
of work (W) and heat (Q) added to the system. [65], [36]

$$\Delta U = W + Q \tag{2.2}$$

Work and heat themself are so called process quantities. They do not describe any state of
a system, in contrary to state quantities, but the path that the system took to reach that
state. [65] *Work* is defined, according to classical mechanics, as the scalar product of a force ($\vec{F}$)
and the movement ($d\vec{r}$) caused by it. [65]

$$W_{12} = \int_1^2 \vec{F} \cdot d\vec{r} \tag{2.3}$$

In a thermodynamic sense there exist other forms of work. Potential, kinetic or electric energy
are considered to be work, if they can add energy to a system in some kind of form [4]. One
of the most relevant forms, for this project is work in the form of expansion work. Expansion
work describes the work that is needed (or can be subtracted) to change the volume of a system.
This, for example, happens during the expansion of the pressurized steam within the expander.
[65], [53]

$$W_{Exp} = \int_{V_1}^{V_2} pdV \tag{2.4}$$

Heat can be defined as energy that is transferred between a system and its environment (or
between two systems) across its system border(s) solely because of a driving temperature difference ΔT. This means, heat is a quantity that can cross system borders just like work. It is
calculated by integrating the heat flow ($\dot{Q}$) over the system border:

$$Q_{12} = \int_{t_1}^{t_2} \dot{Q}(t)dt \tag{2.5}$$

Enthalpy
Enthalpy (H) consists of the internal energy of a system but also takes the product of pressure
and volume of the system into account. It is determined as: [65]

$$dH = d(U + p \cdot V) \tag{2.6}$$

Entropy
Entropy (S) is the third of the main caloric state quantities. It was defined by Clausius (see
chapter 2.1.6) as a criterion for the quality of heat. The higher the temperature of an amount
of heat, the smaller is its entropy [53]. The definition of entropy is:

$$dS = \frac{dQ_{rev}}{T} \tag{2.7}$$

Internal energy, enthalpy and entropy cannot be measured or stated in their absolute amount.
It is only possible to give their amount in relation to a reference point.

2.1.3 Thermal Transfer and Heat Flow

Thermal Transfer
According to Baehr [3], the teachings of thermal transfer declare which influence the temperature gradient has on the heat transfer from a warm object to a cold object and how fast or intense the process happens. The different kinds of thermal energy transportation will be briefly illustrated in the following to provide insight into the teachings of heat transfer.

Thermal conduction
Thermal conduction describes the phenomenon of energy transportation between two adjacent molecules or atoms caused by a temperature difference between them. The energy is conveyed from the particle with higher temperature to the particle with lower temperature. Within radiation-impermeable solid bodies, this is the only form of thermal transfer. Radiation-permeable solid bodies (e.g. glass) can also transport thermal energy within themself by thermal radiation. Fluids can also transport heat by convection besides the ways of thermal conduction and thermal radiation. [3], [7]

Thermal radiation
Thermal energy can be transferred without the dependency on atoms or molecules by radiation. Any body, no matter if solid, liquid or gaseous, emits thermal energy in the form of radiation, if its temperature is above absolute zero. This means that even a colder body radiates onto a warmer body while the warmer body also radiates onto the colder one. The technical interest normally is centered on the net emission, which has to be calculated depending on the absorption, reflection and transmission of both bodies and their environment. Thermal radiation is of lesser importance for the following investigations and is thus not further discussed. [7]

Convection
Technically speaking there are only two kinds of thermal transfer: conduction and radiation. Convection can not happen independently from these two. Yet it is often seen as equal because of its technical relevance. It describes the thermal transfer to or from a flowing fluid. The fluid can be forced to flow by an external pressure difference (forced convection) or it can flow because of a difference in density caused by the increase in temperature (free convection). The thermal transfer through conduction is enforced through the movement of particles which makes convection an effective way of thermal transfer. [3], [7]

Heat Flow
The characteristic quantity of the heat transfer is the heat flow, which possesses a different calculation formula for each kind of heat transfer. Mainly two kinds of heat flows are relevant for this work: the wall heat flow and the heat flow to a medium. The first one describes the amount of heat that is dispatched from a fluid to a wall (or vice versa). The second one is used to describe the impact of that heat flow on the medium.

Wall Heat Flow
According to the Fourier's law for a one dimensional heat conduction the heat flow within a wall (W) is dependent on the area (A), the thermal conductivity of the material (λ), the temperature difference and the distance between two points: [3]

$$\dot{Q}_{W-W} = A \cdot \dot{q}_W = A \cdot \left(-\lambda \cdot \left(\frac{\delta T}{\delta y} \right)_W \right) \tag{2.8}$$

The heat flow between a wall and fluid (Fl) is predetermined by the thermal boundary-layer, which is described by the heat transfer coefficient (α). It has to be calculated for different geometries and fluid velocities. [100]

$$\dot{Q}_{W-Fl} = A \cdot (\alpha \cdot (T_W - T_{Fl})) \tag{2.9}$$

Heat Flow of a Medium

The potential heat flow that is carried by a medium (mass flow $\dot{m}$) can be calculated when it is taken in relation to a reference temperature of a heat source/sink. By setting the temperature of exhaust gas (EG) in relation to ambient temperature it is possible to describe the heat flow of exhaust gas as: [4], [7], [14]

$$\dot{Q}_{EG} = \dot{m}_{EG} \cdot \int_{T_{Amb}}^{T_{EG}} c_{p,EG} \, dT \tag{2.10}$$

With c_p being the specific heat capacity of the fluid, which can be gathered from tables or data bases (normally given as a mean value for a certain temperature range [53]). For a mixture of gases like exhaust gas, the overall specific heat capacity can be determined by calculating the weighted sum of its components. [14], [20]

The heat flow of the working fluid (WF) within the exhaust gas heat exchanger (EGX) can be calculated by using the difference in its specific enthalpy (h) ahead and after the EGX. Therefore changes in pressure and state (vaporization) are considered.

$$\dot{Q}_{EGX} = \dot{H}_{WF} = \dot{m}_{WF} \cdot (h_3 - h_2) \tag{2.11}$$

2.1.4 Efficiency

Efficiency is an important topic for this work. The main purpose of a WHR-system is to improve the overall efficiency of the drivetrain and the whole vehicle. This is achieved by supplying it with additional power. As long as the overall efficiency is increased, it is admissible to decrease the efficiency of the WHR-system. This can for example be achieved by using additional heat sources (cf. chapter 4.2), as they could technically decrease the efficiency of the WHR-system while at the same time increasing the power output. In this case the overall system efficiency is increased, while the efficiency of the WHR-system is decreased.

The standard definition of efficiency (η) is the ratio between the usable power output (P) of a system and the power that was put into it.

$$\eta = \frac{P_{out}}{P_{in}} \tag{2.12}$$

Complex systems need a more detailed definition of their efficiency, as it is not always obvious what in and out going power, as well as usable power, are. Thus the different kinds will be defined here.

The efficiency of a heat exchanger (HX) is defined as the ratio of heat flow it is able to transport from its heat source to its heat sink and the amount of available heat flow, which is the exergy flow of the exhaust gas. For an EGX the formula looks like equation 2.13.

$$\eta_{EGX} = \frac{\dot{Q}_{EGX}}{\dot{H}_{EG}} \tag{2.13}$$

The efficiency of WHR-systems can be defined as the ratio of heat flow that is transferred into the system (in this case with a single EGX) versus the power it is able to provide. The power can be provided in form of electricity or mechanically depending on the system. For the system that is used in this thesis, the power output is directly forwarded to a generator which converts it into electric energy. Thus any generating losses are attributed to the system. The formula looks like equation 2.14.

$$\eta_{WHR-system} = \frac{P_{\text{WHR}}}{\dot{Q}_{EGX}} = \frac{P_{el}}{\dot{Q}_{EGX}} \tag{2.14}$$

If there is more than one heat source, they all have to be included.

$$\eta_{WHR-system,overall} = \frac{P_{WHR}}{\dot{Q}_{EGX} + \sum \dot{Q}_{HX}} \tag{2.15}$$

When the efficiencies of all heat exchangers are also included, the formula changes to equation 2.16.

$$\eta_{WHR-system,absolute} = \frac{P_{\text{WHR}}}{\dot{Q}_{EG} + \sum \dot{Q}_{OtherHeatFlows}} \tag{2.16}$$

With an imaginary line drawn around the whole car it is possible to define the overall system efficiency of the car. Outputs are effective power output from the engine and power output of the WHR-system. The only input would be the power of the fuel. This results in equation 2.17.

$$\eta_{System} = \frac{P_{\text{WHR}} + P_e}{\dot{m}_{Fuel} \cdot H_u} \tag{2.17}$$

Another efficiency has to be defined to describe the advantages of the additional power generation through altering the engine application which is described in chapter 4.3 and to describe other system influencing parameters. Here the additional power output of the system through an alteration is compared to the power of the additional fuel consumption.

$$\eta_{Alteration} = \frac{\Delta P_{\text{WHR}}}{\Delta P_{Fuel}} \tag{2.18}$$

This value can theoretically reach values above 100 % as a measure can improve the power output of the system while being almost neutral to the engine's fuel consumption (e.g. adding a recuperator). The result alone does not represent a meaningful benchmark for an alteration. Furthermore the directions of ΔP_{WHR} and ΔP_{Fuel} have to be taken into account:

If the value of $\eta_{Alteration}$ is negative because ΔP_{WHR} is negative and ΔP_{Fuel} is positive, the alteration is pointless.

If the value of $\eta_{Alteration}$ is negative because ΔP_{WHR} is positive and ΔP_{Fuel} is negative, the alteration is without any disadvantage and thus beneficial.

If the value of $\eta_{Alteration}$ is positive because ΔP_{WHR} is negative and ΔP_{Fuel} is negative, the alteration is technically beneficial for values up to 100 %. Above that value the power output of the WHR-system is reduced by a larger amount than fuel power is saved.

If the value of $\eta_{Alteration}$ is positive because ΔP_{WHR} is positive and ΔP_{Fuel} is positive, the alteration is technically beneficial for values above 100 %. When the WHR-system is compared to the engine the alteration can be considered beneficial as long as the value is above the efficiency of the engine (for the engine used in this project around 20-35 %). Otherwise the engine would use the fuel more efficiently.

2.1.5 Thermodynamic Cycles

This subchapter give an overview of the most relevant thermodynamic cycles, that are either used as comparison (Carnot cycle) or used in similar ways as a Rankine cycle in other applications (Otto and Brayton cycle). A principle depiction of all cycles can be found in appendix A.2.

Carnot Cycle

The Carnot cycle is a theoretical cycle that is commonly used as a comparison to the maximal achievable efficiency for a given heat sink and heat source. The Carnot-efficiency is the maximal efficiency that a thermodynamic engine can reach. It is only dependent on the temperature of the heat sink (cold) and the heat source (hot). As these values can not reach 0 or ∞, the Carnot-efficiency cannot reach a value of 1. [100]

$$\eta_{Carnot} = 1 - \frac{T_{Cold}}{T_{Hot}} \tag{2.19}$$

The theoretical cycle consists of an isothermal compression, an isentropic compression, an isothermal expansion and an isentropic expansion. For the best possible efficiency the heat addition should be executed at the highest possible temperature and the heat dissipation at the lowest possible temperature. The compression and the expansion should occur with as little increase in entropy as possible. [20]

Otto Cycle

The Otto cycle is used to describe the processes happening during the combustion in a spark ignition (SI) engine. The process includes an isentropic compression (compression stroke), an isochoric heat addition (ignition and combustion), an isentropic expansion (expansion stroke) and isochoric heat rejection (charge exchange). This cycle is, as most thermodynamic cycles, only an approximation to reality. There are no isentropic processes as each process is prone to friction and heat loss; also the heat addition and rejection can not happen instantly ($\Delta V \neq 0$). Yet this cycle can be used to explain the thermodynamic coherences taking place in an engine. [65]

Brayton Cycle

The Brayton cycle or Joule cycle is a thermodynamic cycle that is used as a comparison model for the processes happening within gas turbines. The changes of state consist of an isentropic compression (compressor), an isobaric heating (combustion of fuel or heat exchanger), an isentropic expansion (turbine) and an isobaric heat rejection (exhaust or cooler). While the cycle schematic is similar to the Rankine cycle (see figure 2.2) the working fluid is not. The working fluid can be a gas that is recirculated, or if the cycle-design is open, it can be ambient air that is used and combusted with fuel. [94]

2.1.6 Clausius Rankine Cycle

The Clausius Rankine cycle (often only Rankine cycle) serves as a comparison model for steam cycles used in thermal power stations. It is used where ever heat is available at elevated levels and amounts, like in power plants, ships or solar thermal power stations. The cycle was developed by William John Macquorn Rankine, a Scottish mechanical engineer. Also attached to the cycle is Rudolf Julius Emanuel Clausius, a German physicist. Both are considered to be founding fathers of the field of thermodynamics.

The Rankine cycle is an idealized thermodynamic cycle in which heat is converted into mechanical energy by the use of a working fluid that is vaporized by a heat source and expanded within an expander. The cycle is closed by condensing the remaining fluid with a heat sink and re-pressurizing the fluid with a pump. The usual applications are coal or nuclear power plants where the cycle is used in advanced form to generate electricity. The common working fluid used therein is water. The heat from the burning coal or from the nuclear reactor vaporizes the water. The resulting steam drives a turbine which produces mechanical energy that is converted by a generator. The process generates energy due to the fact that a quasi incompressible fluid is pumped and a gaseous steam is expanded, the existing difference in volume flow, pressure and temperature propels a turbine. [100]

To operate a Rankine cycle at least four components have to be provided: a pump which transports the working fluid through the system, a boiler which vaporizes the working fluid, an expander which transforms the steam energy into usable mechanical energy and a condenser which cools down the working fluid until it is condensed. Figure 2.1 shows an example for a cycle in a T-s-diagram.

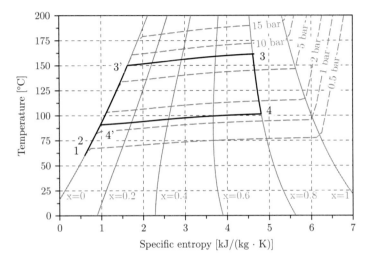

Figure 2.1: Example of a T-s-diagram for a Rankine cycle with wet steam, wet working fluid and without superheating. The left line at x = 0 is called the saturated liquid line and the one with x = 1 is the saturated vapor line.

Changes of State

1 – 2: Isentropic compression: The working fluid is pumped into the circuit against the back-pressure of the boiler and the expander, which results in an increase of pressure. The power input is low due to the liquid state of the fluid. Thus temperature and entropy aren't visibly different.

2 – 3: Isobaric heating: Within the boiler the working fluid is first heated up (2-3') and then vaporized (3'-3) up to the desired steam quality (x), which in this case is 0.8 (wet steam). Due to the working fluid being a zeotropic mixture, the vaporization is not isothermal and thus the steam quality can be directly calculated through measuring the values of pressure and temperature after the boiler. If the expander is a turbine the steam is normally superheated (x > 1).

3 – 4: Isentropic expansion: From point 3 to 4 the steam is expanded within the expander. Temperature and pressure of the steam are reduced; the energy of the steam is used to propel the expander. The resulting mechanical energy can be extracted at the shaft of the expander.

4 – 1: Isobaric cooling: After the expansion the remaining steam is condensed in order for the pump to provide the required mass flow efficiently and reliably. This is done by the condenser. The condenser uses a low temperature heat sink to first cool down the steam (4-4') and then condense and undercool it.

The steam quality is defined as the ratio of the mass fraction that is in steam form to the mass of the overall fluid:

$$x = \frac{m_{Steam}}{m_{Overall}} = \frac{m_{Steam}}{m_{Liquid} + m_{Steam}} \tag{2.20}$$

By this definition saturated steam has a quality of 1 (or 100 %) and saturated liquid has a quality of 0 (or 0 %). Values in between indicate that the steam is "wet" - fluid is left that is not vaporized. There are also values below 0 (for sub-cooled fluid) and above 1 (for superheated or "dry" steam) possible. These values describe how much more (or less) enthalpy the fluid has, than it would have in its saturated states. A definition is given in formula 2.21, wherein h is the current enthalpy, h' the enthalpy of saturated liquid and h'' the enthalpy of saturated vapor. [53]

$$x = \frac{h - h'}{h'' - h'} \tag{2.21}$$

The schematic of the basic cycle is depicted in figure 2.2. The pump puts the power P_{12} into the fluid which increases its pressure. Within the boiler the heat flow from the source $\dot{Q}_{23}$ rises its temperature and vaporizes the liquid fluid to steam. The expander transforms the energy of the pressurized steam into mechanical Power P_{34} and the remaining steam is condensed under the dissipation of the condenser heat flow $\dot{Q}_{41}$.

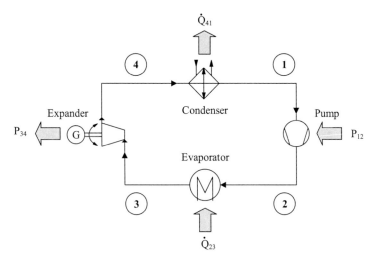

Figure 2.2: Schematic depiction of the basic cycle including the in- and outputs of heat and power

The working fluid for a Rankine cycle can be any fluid that can be vaporized by the given heat source. Working fluids are a vast field of research on their own and thus will not be discussed in detail in this work. Further details can be found in the work of Körner [57] and Preißinger [82].

Efficiency Improvement

There are several ways to improve the efficiency of a Rankine cycle outside of the obvious ways: enhancing the efficiency of the pump and the expander. There are multiple ways to increase the thermodynamic efficiency by advancing the cycle design and the operating conditions.

According to Carnots theorem the maximal efficiency of a cycle is dependent on the high- and low-temperature level of it. For a Rankine cycle this translates to the demand that the steam temperature should be as high as possible. In coal fired power plants the steam temperature reaches about 650°C [17]. Higher temperatures would be possible for the heat source, but are limited by the thermal stability used in the materials of heat exchangers and piping [20]. The low-temperature is given by the heat sink, which is normally water from rivers (e.g. max. 20-25°C by law in Baden-Württemberg [62]) or cooling towers (thus ambient air). These sinks impose the minimal temperature. Lower temperature sources managing these amounts of heat are not permanently and efficiently feasible.

An improvement through a different kind of cycle design can be achieved, e.g. by reheating the steam. By adding heat to the steam after the first stage of expansion and thereby increasing the quality of the steam, another lower pressure expansion is made possible without the risk of damaging the turbine through condensation of the steam. In modern power plants even double reheating is common. This design requires additional piping, an additional boiler and multiple turbine stages in series, but makes better use of the gathered thermal energy and improves the turbine efficiency. [17], [53]

The most common alteration of the Rankine cycle is the regenerative Rankine cycle. The working fluid after the pump is heated by hot working fluid that has already been expanded. This can either be achieved by using a recuperator or by directly using hot steam from the expansion process. The benefit of heating the working fluid after the pump is that it is brought closer to the saturated state. Thereby less heat is added at lower temperature levels and more heat of the source is used to vaporize and superheat the fluid, which increases the efficiency of the cycle.

Another point is the reduction of the condenser heat loss and volume flow. The heat does not have to be dissipated by the condenser and is instead transferred to the cold working fluid. [17]

Organic Rankine Cycle

Another kind of Rankine cycle is the "Organic Rankine Cycle" (ORC). The process of the ORC is basically the same as the one of the Rankine cycle. The difference is mainly found in the utilized working fluid, which is an organic working fluid with a lower boiling temperature than water. The organic fluid can be for example: n-pentane, toluene, ethanol or silicon oils [87]. Applications are found in fields where low temperature heat sources are found that offer large amounts of heat like: solar ponds, geothermal power stations or waste combustion. Due to the lower temperature level of the heat source the ORC is thermodynamically limited in its efficiency. Yet it makes gathering the energy possible and can reach higher efficiency at the heat transfer. Rankine did not prescribe any working fluid, thus the ORC can be considered to be "just" a Rankine cycle. Yet because of the need for matching components, the ORC is often used as name for any application other than water based ones. [43]

Kalina Cycle

The Kalina cycle is a variation of the Rankine cycle. Developed by Alexander Kalina in the 1970s it is supposed to be able to efficiently use low-temperature heat sources. This is achieved by using a working fluid that consists of two components with different boiling temperatures. In most applications a mixture of water and ammonia is chosen. The cycle makes use of the zeotropic properties of the mixture by changing the concentration at the boiler and the condenser to achieve higher mean temperature differences. The improved heat in- and output increases the efficiency. Yet the cycle needs a separator and additional piping and throttles, making the layout much more complex. [43], [109]

2.2 State of the Art in Mobile Applications

This chapter is supposed to describe the current state of the technology of waste heat recovery in mobile use, the limitations that a Rankine cycle based system has to deal with and the boundary conditions that are dictated by mobile applications.

2.2.1 Waste Heat Recovery Technologies

Several technologies are currently in development, intended for use in mobile applications. The amount of heat that is dispatched in a car's exhaust and the regulatory CO_2-limits are the driver for this battle of technologies. Comparisons have been carried out by Dawidziak [14], Armstead [2] and Tauveron [104] mostly showing that each technology has its own benefits and disadvantages. Some of them like the direct use or the turbo-compound have already found their way into production for their most ideal application, but none of them has reached vast deployment in mobile applications. The more complex ones have not reached the final stage of refinement, that is needed for the implementation in a car and are thus not in series production, despite their potential. The most common and promising ones will be discussed in comparison to the Rankine cycle in the following.

Thermoelectric Generator

A thermoelectric generator (TEG) is able to directly convert thermal energy into electric energy without prior transformation into mechanical energy. The TEG is based on the Seebeck-effect:

When two different conductive materials are connected to a heat source and a heat sink an electric potential is induced. Power can be withdrawn depending on the temperature difference and the materials. The choice of material is crucial for its efficiency and makes the TEG technically challenging despite its lack of moving parts. [40]

A potential synergy in the additional cooling of the exhaust gas ahead of the turbocharger with a TEG was investigated by Risse [85]. Both TEG and the reduced fuel enrichment in higher loads reduced fuel consumption, but the additional thermal mass had a drawback on the light-off of the catalyst. BMW [68] and DLR [30] cooperated and built a working TEG into a passenger car. The system was able to generate 200 W at a car speed of 135 km/h with an efficiency of 2 %, which is a great deal smaller than BMW predicts for its Rankine-system [29].

Direct Use

The direct use of exhaust gas heat is not a new approach, the waste heat is not converted, but rather transferred or stored in a defined place. Even the VW Beetle had a kind of heat exchanger added to its exhaust pipes to provide its passengers with heated air, which he otherwise could not have done due to his lack of a coolant system [42]. Newer approaches use the exhaust heat to increase the coolant temperature during the cold start of the engine. This improves the fuel economy and the comfort in the passenger cabin. In order to transfer heat from the exhaust gas to the coolant, a heat exchanger is put into the exhaust system. Several different designs exist for this purpose.

Audi [21] developed a system consisting of an extended cylinder head geometry at the outlet side of the engine with an integrated heat exchanger. This design has two advantages: first the heat input in cold start and second the exhaust gas cooling at higher loads which allows for reducing the fuel enrichment. Other designs put the heat exchanger at a position after the aftertreatment system to avoid influences. They use a bypass pipe to lower the heat input after cold start which takes load off the cooling system. Commercial examples and their benefits are presented by Behr [44] and Faurecia [12]. Hepke [45] investigated the direct heating of different components of the car including automatic transmission, rear axle transmission and cabin heater. He found that all versions can improve the cold start behavior of a car but to varying degrees and even suggested to use the indirect heat from other waste heat technologies like TEG.

Thermoacoustic Generator

Thermoacoustic generators (TACG) make use of the thermoacoustic effect. A sound wave is amplified in the presence of a temperature gradient and the resulting standing-wave or travelling-wave can be transformed into usable kinetic energy by a Helmholtz resonator couple with a linear generator or a bidirectional turbine. Their advantage is the small amount of moving parts and the relatively low cost of its components. They have not yet reached a higher degree of development and only a small number of prototypes has been constructed so far.

Kruse [59] explained the function and showed first results in his work. These show promising efficiencies and power outputs, but can not reach the levels of a Rankine-system. The prototype is also too large and heavy for a mobile application in a car.

Sahoo [88] presented a TACG-system for the use in a commercial truck that reached a reasonable size for this kind of application. His simulations predicted power outputs in the order of magnitude of a comparable Rankine-system. Yet a proof-of-concept with actual hardware has to be performed to validate these values.

Steam-Jet Refrigeration System

This approach uses the thermal energy of the exhaust gases to provide cooling for parts of the car. It can be used to cool the cabin or to lower the charger air temperature, which generates benefits in terms of emissions, risk of knocking, maximal charge mass and efficiency as Ramsberger proved [83]. A pressurized working fluid is vaporized within an EGX and the resulting large high-pressure volume flow is then used to create a suction pressure. This is achieved with an ejector nozzle that transforms the potential pressure energy of the hot steam into kinetic energy. A low pressure within the nozzle is the result. Cold working fluid from the low-temperature part of the cycle can be sucked in by that low-pressure. With the use of a throttle this can be used to vaporize the working fluid. The resulting drop in temperature can be utilized in a cooler to bring any fluid (cabin air, charge air,...) to temperatures even below the ambient temperature. The power consumption of this system is lower than the one of a refrigerant circuit as most of the applied energy is delivered by the exhaust gas.

Kadunic [51], [52] was one of the first to create a system based on the steam-jet refrigeration and demonstrated the function on an engine test bench.

Thoma [105] presented a system designed for a passenger car that uses the exhaust heat to cool down the charge air. He showed that the use of recuperators is beneficial for the cooling power and for the amount of heat that is dissipated by the condenser.

Turbo-Compound

Exhaust gas turbochargers themselves are a kind of waste heat recovery. The enthalpy flow of the exhaust gas and its kinetic energy are used by a turbine to drive a compressor. The compressor increases the pressure of the charge air and thus the energy of the exhaust gas is transferred back into the system. A further development of that concept is the turbo-compound (TC). An exhaust turbine is used to propel a transmission that produces either usable mechanical energy or is coupled to a generator to provide electric energy. The turbine can either be added after the turbocharger or the turbine of the turbocharger itself can be used.

Dawidziak [14] compared the turbo-compound to other WHR-systems for the use in motorsports. He came to the conclusion that the TC is the most promising technology due to its small impact on weight and relatively high power output. The additional possibility to support the pressure built-up was another important point for this application. Yet he found Rankine-systems to have a higher potential power output.

Günther [39] tried to characterize a turbo-compound system in comparison to an asymmetric turbocharger on a commercial diesel engine. He predicted benefits in fuel consumption despite the raised exhaust gas back-pressure but suspected disadvantages in terms of exhaust gas recirculation. His simulation results also indicate a slight disadvantage for a downstream integrated ORC-system due to the lowered exhaust gas temperature and mass flow, but pointed out that an ORC-system would still be advantageous.

2.2.2 Limitations of the Real Clausius Rankine Cycle

The changes of state presented in chapter 2.1.6 do not absolutely reproduce the changes of state of a real Rankine-system. A real Rankine-system has to work under non-ideal conditions, which means every change of state underlies irreversibility due to heat loss and friction. The detailed limitations for each change of state are presented in the following.

Pressurization

In reality the compression by the pump is not isentropic. The pump produces entropy caused by friction between its moving parts and within the fluid itself. The pressure difference between high- and low-pressure leads to leakage through crevices of the pump. Both friction and leakage increase the power consumption of the pump. The losses of the pump attribute to some extend to the working fluid in the form of heat and can therefore be regained to a small portion.

Vaporization

The vaporization is not isobaric, as the transportation of the fluid and especially the transportation of the multiple times larger steam volume flow, produce large pressure losses. These result in lower pressures after the vaporization. To reach the same pressure in the end, the pump has to provide higher pressures, which adds to its power consumption.

Condensation

Just as the vaporization, the condensation is subject to pressure loss. The lowest pressure within the system is found before the pump, this means any pressure loss over the condenser adds to the pressure level after the expander. This increases the temperature level after the expansion and reduces the potential recoverable energy.

Expansion

The expansion in the turbine is not isentropic either. Friction within the expander reduces the power output of the expander and increases the entropy. The temperature after the expansion is higher than it theoretically could be due to leakage, friction and non adiabatic conditions [57]. This further reduces the potential energy that can be regained.

Expansion is also limited by the steam quality. The lower the steam quality the more water is condensed during the expansion and the formation of water droplets is more likely. These droplets can lead to erosion and pitting on turbine blades and housing. Over time this inflicts damage on the turbine and reduces its efficiency. Reheating and superheating can only avoid but not disable this effect. In literature minimally exceptable values of 0.85 can be found for the steam quality after their expansion [17]. If an expander is used that does not need superheated steam, then the limit of steam quality is lower. A damaging effect can be ruled out due to the lower speeds of the moving parts.

For rotary positive displacement expansion machines, like scroll expanders, the topic of optimal pressure ratio is highly significant, especially for mobile applications with constantly changing heat inputs. The pressure ratio is defined as the ratio between the pressure of the fluid at the inlet of the expander and the pressure at the outlet of it: $\Pi = p_{in}/p_{out}$. Due to the design of these machines the geometric expansion ratio ($\epsilon = V_{out}/V_{in}$) and thus the (optimal) pressure ratio are fixed (in first approximation). Any alteration from the optimal pressure ratio leads to over- or under-expansion of the steam. These have disadvantages on the efficiency of the expander.

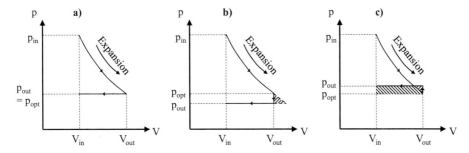

Figure 2.3: Different kinds of expansions resulting from different pressure ratios for the same level of high-pressure (p_{in}) and the same geometric expansion ratio. p_{out} ($\neq$ const.) is the pressure outside of the outlet of the expander and p_{opt} is the pressure of the fluid after the expansion within the expander. a) optimal expansion, b) under-expansion, c) over-expansion. Shaded areas represent thermodynamic losses

Figure 2.3 shows an idealized depiction[1] of the influence of the pressure ratio. The high-pressure is kept constant and the low-pressure varies, which would translate to fixed high-temperatures and varying low-temperatures for a given steam quality. This would mean different potential Carnot efficiencies. Yet this depiction allows for a simple depiction of the thermodynamic losses. The optimal pressure ratio ($\Pi = \Pi_{opt}$) as depicted in part a) gives an expansion ending at a pressure that is as high as the pressure at the outlet $p_{out} = p_{opt}$. No potential energy is wasted, when the expansion is finished and the outlet port opens. In the case of under-expansion b) the steam is not expanded down to the outlet pressure level: $\Pi > \Pi_{opt}$ and $p_{out} < p_{opt}$. The potential of the pressure ratio is not used entirely by the expansion and the steam expands further when the outlet is opened. Over-expansion c) occurs when the steam is expanded to a pressure level lower than the one of the outlet $\Pi < \Pi_{opt}$ and $p_{out} > p_{opt}$. The opening of the outlet results in a pressure increase and the steam has to be ejected against that pressure. Over-expansion has a worse influence on power output than under-expansion due to the expansion work that is put into the fluid and the additional ejection work. [13], [67]

The ideal pressure ratio can either be achieved by setting the low-pressure of a system to a suitable value or by adjusting the high-pressure. The latter can be implemented by opening more or less flow nozzles in turbines. For scroll expanders this can be done by adjusting the outlet position of the scroll (further out or in) which changes the expansion ratio. [112], [77]

2.2.3 Boundary Conditions

The purpose of this subchapter is to give an introduction of the conditions that a WHR-system installed in a car has to be able to cope with. These boundary conditions have an influence on the system design, choice of materials, the cycle design and most importantly on the possible efficiency and power output of the WHR-system.

Restrictions due to Mobile Applications
A WHR-system has to work with certain special restrictions due to the mobile application in a car. These restrictions come on top of the thermodynamic limitations for a real cycle. General restrictions of any system that is implemented in a car come from the limitations of weight, packaging and environment. Systems have to be light to have a small impact on the fuel economy and the dynamic of the car. They have to be small to be able to be fitted into the limited space available in a car.

The environment of a car can be very different in terms of temperature, pressure and humidity. These differ from region to region and also differ between summer and winter. The WHR-system has to be able to work at or at least withstand any of these general restrictions to be considered for deployment.

The engine is the main heat source of a car. Heat can be drawn most efficiently from its coolant system and from its exhaust gas system thanks to their relatively high temperature levels. The amounts vary depending on the operation point of the car. Generally the exhaust gas has the highest heat flow and the highest temperature levels. These can differ depending on the engine and its load and speed. For the engine of a mid-size passenger car (like the one used in this project) the values lie between 2 kW - 180 kW and 370°C - 920°C after the catalyst. The wide spectrum is a characteristic of the utilization in a car. Cars are exposed to different driving behaviors and cycles that the engine is delivering exhaust gas heat for. A variety of operating points must be taken into account. For example: low load city commute to high speed highway

---

[1]A similar depiction can be found in the work of Körner [57], but there the geometric expansion ratio is altered and not the pressure ratio.

driving, pulling loads up steep hills, dynamic driving with quick load steps and long times without any operation at all.

As the exhaust gas of the engine is the most common heat source for a WHR-system, it will most likely have to be placed somewhere along its exhaust system. The exhaust system is exposed to high ambient temperatures, dirt, potential collision with the ground or debris, vibration and thermal stress. These dictate some security measures on the WHR-system to ensure save operation.

The main purpose of the exhaust system is the provision of charge air pressure by the turbocharger, the aftertreatment of the exhaust gas and the modification of exhaust sound. The clear priority on driving experience through quick torque buildup and the regulatory requirements to reduce exhaust emissions define the hierarchy along the exhaust: first the turbocharger then the three-way-catalyst and afterwards the WHR-system. Any higher ranking of the WHR-system would inflict the performance of the other two. With the possible future deployment of gasoline particulate filters (GPF) this position might even be reduced to number four. A conflict with the noise of the exhaust system cannot be generally stated, but the possibility to omit one of the mufflers is open [42].

There are two possible heat sinks that are available in a car: the ambient air and the coolant circuit. If ambient air is used, the condenser would have to be designed to be directly flownthrough by the airflow that results from the cars movement. This means a radiator has to be added to the car that has a connection to the environment and possibly needs a fan to improve its heat dissipation. The disadvantage in aerodynamics and the power consumption of the fan can be significant as Erlandsson [23] presented in his work.

The other heat sink being the coolant circuit of the car, which is normally designed for the maximal engine heat output at low car speeds which occur for example when pulling a load up a hill. As this only happens at a very small portion of the possible engine operating points, most of them posses a capital cooling reserve that can be utilized [42], [46]. But this reserve gets smaller the higher the engine load is and higher engine loads mean higher heat inputs from the WHR-system. Yet the reserve is expected to be able to cover the heat load from a WHR-system up to medium engine loads (upto 150 km/ or 50 kW engine power [46]). A bypass system for the EGX has to be applied to prevent overloading of the cooling system at higher engine loads.

The power generated by the system has to be used or stored by the car. If the power is generated in mechanical form it can be transferred to the drivetrain by a transmission or a clutch depending on the rotational speed of the expander. This restricts the position of the expander and comes with additional friction losses. Also the direct impact on the engine load results in a feedback on the exhaust heat flow and thus power output and so on.

The generation of electric power grants more freedom in placement of the expander and in the use of the power. The future development of 48 V electrical systems and the increasing hybridisation of passenger cars make this approach more promising [77]. The battery capacity and the general consumption of electric power will be appropriate to the potential electric output of a WHR-system. The topic will be discussed in greater detail in chapter 2.2.4.

Restrictions to the System Components

A WHR system that is working on the principles of Rankine and that is limited to the surrounding of an automobile typically has to consist of at least five specific components. Figure 2.4 shows those components and their connections to the car. The pump is responsible for transporting the working fluid through the system and delivering the needed pressure. The exhaust gas heat exchanger transfers the heat from the exhaust gas to the working fluid. It is vaporized within the heat exchanger and the resulting high pressure steam is used to propel the expander. The expander converts the enthalpy flow of the steam into mechanical energy on a crank shaft where it can be used for driving the car or producing electric energy. The remaining steam is condensed by the condenser and the change in volume of the fluid is compensated by an expansion reservoir. [33]

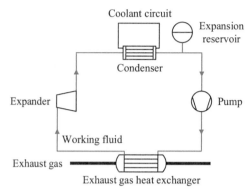

Figure 2.4: Schematic of a basic Rankine-system for automotive use

The working fluid is the focal point of the system design. The working fluid influences the choices of the expander principle, the design of the EGX, the size of the condenser, the low-pressure of the system and the size of the pump. Thus it is selected before any other component is designed. It is chosen based on its properties of vaporization temperature/pressure, condensation temperature/pressure, enthalpy progression, freezing point, kinematic viscosity, thermal stability, cost, and harmfulness to the environment and health [87]. Generally the form and position of its saturation dome is most important. Fluids with a falling saturated liquid line are called "wet", ones with a vertical line are called "isentropic" and a fluid with a ascending line is described as "dry". Possible fluids can be, but are not limited to: R134a, water, ethanol etc. or mixtures of those. An elaborate comparison can be found in the works of Körner [57] and Heberle [43]. Ethanol is often seen as the most promising fluid due to its performance, low global warming potential and cost despite its ability of being flammable [92], [82].
There is often an addition of some kind of oil for lubrication purposes. This oil can be necessary but also brings negative thermodynamic aspects with it. The oil is heated and cooled within the circuit, but does not deliver any steam and thus reduces the power output and efficiency. Also it was found that the oil separated from the working fluid during the vaporization (if it does not vaporize itself) and is transported along the walls of the boiler where it hinders the heat flow, or might even decompose [103]. This could be avoided by installing an oil separator, recycling the oil to a point where it is needed. [80]

The pump of the system has to be able to provide a wide variety of mass flows and pressures. It should deliver this with a high efficiency as the power consumption of it is in direct conflict with the power output of the system. The most common pump design is a positive displacement pump, as they are able to efficiently deliver the needed pressure and can be adapted in size to realize the mass flow range [99]. Two kinds of drives can be chosen with reasonable complexity. Usually the pump is driven by an electric motor which allows the highest freedom in terms of placement and speed but the electric power consumption has to be provided by the electric system which can be troublesome with a 12 V system as the consumption can reach multiple hundreds of watts. The other way is to use the expander shaft as a mechanical drive. This couples the speed of expander and the pump, but simplifies the system [48]. In this configuration mass flow cannot be adjusted freely to accommodate the given heat source which results in efficiency losses.

Restrictions of the expander are mostly due to the working fluid and the steam quality. While some positive displacement expansion machines (like scroll expanders) can work with wet steam (or even are more efficient) others like turbines need superheated steam. Also dependent on the kind of expansion machine is the lubrication. While some principles are self-lubricant (e.g. rotary vane expanders), some are able to get along with liquid fluid and others need the addition of a lubricant (e.g. silicon oil) [80]. The rotational speed is another characteristic of an expansion machine. Turbines revolve with speeds around 70.000 rpm [14] while a piston expander will work in the lower thousands [9]. This has to be considered when utilizing the mechanical power by a generator or by the engines drivetrain. Detailed information on this topic can be found in the FVV final report "Expansionsmaschine" [95].

The most common design for an EGX is a plate heat exchanger as this design allows high amounts of heat transferring surface with small parts and low production costs. The sheets are connected by brazing. The filler material has to remain solid at the high temperatures of the exhaust gas (max. 750°C in this work) and has to compensate the thermal expansion of the sheets. Constant changes in engine load and the resulting changes in temperature level of the exhaust gas as well as the sudden drop in heat input when the system is switched to bypass operation put the exhaust gas heat exchanger under significant thermal stress. While the temperature difference between fluid (max. 180°C in this work) and exhaust gas is small in the back of the EGX, it is huge in the front of it (countercurrent flow), the stress is highest at the entrance of the exhaust gas.

The EGX also has to provide uniform heat flow and working fluid mass flow over all of its sheets. Non-uniformity would result in a difference in steam quality which would result in different pressure drops (Ledinegg instability) [5]. This would intensify the uneven distribution of mass flow and thus result in even higher differences in steam quality. In the end some sheets would produce too superheated steam and some would produce too wet steam. This can even happen without noticing, as the mass flows would merge at the end of the EGX. The downsides to this occurrence are lowered overall heat flows and mostly the danger of local thermal decomposition of the working fluid and/or lubricant.

The exhaust gas side of the EGX is less complicated as the only other criterion is the back-pressure of it, which can be considered to be manageable for most EGX sizes especially when higher exhaust volume flows are handled by the bypass.

The expansion of the working fluid that is liquid inside of the expander and the EGX before the system is started, is compensated by the expansion reservoir. Without it, the low-pressure would drastically climb through the vaporization until an equilibrium of pressure and vapor amount is adjusted (or the system bursts). To keep the low-pressure within the desired range an appropriate amount of space has to be provided by the expansion reservoir. The fluid can be separated by some kind of membrane to protect the system from any kind of gas intrusion or to keep the fluid from evaporating into the environment. The pressure from the environment or gas side of the expansion reservoir is conveyed to the system and adjusts its low-pressure.

The condenser is the most simple component of the system. It is most likely designed as a plate heat exchanger for the same reasons as the EGX. Yet it does not have to cope with the same amount of thermal stress as the EGX. The temperature difference and the temperature level of its fluid are much lower. The main focus for it is the reliable condensation of the steam with as low back-pressure to the expander as possible.

Each component is connected to the other ones by some kind of piping. The piping has to withstand the pressures, temperatures and chemical properties of the working fluid. The piping has to provide thermal insulation, tightness and low-pressure loss with as small as possible impact on size and weight, while at the same time compensating vibrations and thermal expansion.

Restrictions to System Operating Parameters

System operating parameters are defined in this work to be parameters that can be altered while the system is running. They stand in contrast to the design parameter of geometry and boundary conditions. The ones that can be found in a basic system design (and thus must be part of complexer systems) are depicted in figure 2.5.

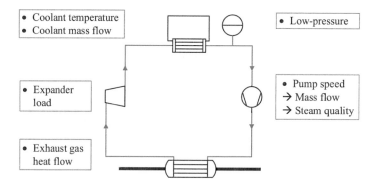

Figure 2.5: Factors of influence on Rankine-systems for automotive use.

The first parameter is the exhaust gas heat flow. This parameter is predefined by the engine operating point which is related to the speed and operation of the car and can differ in amounts of mass flow and temperature as mentioned earlier. In an add-on system this parameter is normally considered to be a given parameter, but as the results from chapter 4.3 will show the heat flow can be changed independent of the cars operation.

The speed of the pump is the most important parameter of the whole system. It is directly responsible for the mass flow within the system. Higher speeds produce higher mass flows and for a constant heat supply from the exhaust gas, they translate into lower steam qualities. The back-pressure from the expander and the exhaust gas heat exchanger are a result of the vaporization and the pressure losses within these components. The pressure has to be overcome by the pump. Normally there is an ideal or required steam quality for a given operating point and expander type, thus the pump speed can not be changed independently. The impact on steam quality and system efficiency will be presented in 4.1.1.

The back-pressure can be partially influenced by the load that is put on the expander. In this work the load was set using different electrical resistances that dissipated the generated power. Higher loads result in lower expander speeds and higher back-pressures, but might not result in higher power outputs. For each engine operation point and steam quality there is one optimal load for the expander at which the maximal amount of power can be collected. This can be explained through the interaction of friction loss within the expander at higher speeds and the thermodynamic loss in efficiency at lower high-pressures / high-temperatures that occur at lower speeds. The loads for the experiments in this project were chosen to deliver high power output at their respective engine operating points and were kept constant for all variations to avoid any unwanted side effects.

The cooling of the condenser can be seen as one parameter in itself. If a coolant condenser is utilized, the mass flow of the coolant and the coolant inlet temperature are two parts contributing to its cooling ability. They determine the low-temperature level within the WHR-system, which is relevant for the possible low-pressure of the system and the Carnot efficiency of the system. The temperature level of the coolant can vary over a huge temperature range. In a cold start the coolant might have a temperature far below 0°C while they normally are located in the range of 40-80°C at steady-state operation. The temperature level can be set to a desired amount by the use of a thermostat. The mass flow can be set to any desired amount with the right choice of coolant pump. Constant mass flows are possible if they are designed for the maximal heat output of the condenser, yet they are not reasonable in terms of efficiency. With a radiator condenser the cooling effect is directly dependent on the ambient temperature and the vehicle speed ($\rightarrow$ air mass flow). They thus lack a degree of freedom, if no air throttle is applied [71]. A detailed analysis of the influence of the condenser cooling will be presented in chapter 4.1.2.

One parameter that is often considered to be constant for mobile applications is the low-pressure. The low-pressure can be set through the pressure on the gas-side of the expansion reservoir. If this side is connected to the environment the pressure is indeed predetermined by the ambient pressure ($\approx$ 1 bar), however, it is possible to adjust that pressure. This can be achieved either by closing the gas-side or connecting an air-pressure reservoir to it. In the first case the pressure is set once at the beginning when the reservoir is closed and after that an equilibrium of gas pressure and fluid pressure from the vaporized fluid adjusts itself accordingly. Smart design of the expansion reservoir and the right choice of gas pressure are essential for systems like that. In the second case an air-compressor is connected to the gas-side that delivers the desired pressure. Both designs can allow pressures below and above ambient pressure, at any desired pressure level. The influence on the WHR-system is discussed in chapter 4.1.3.

Comparison to Rankine Cycles in Power Plants
The Rankine cycle has been used in power plants for a long time and has reached a high level of sophistication and efficiency. The modern forms of the Rankine cycle in power plants are much more complicated than the basic principle and they do have many more stages including superheating, reheating, economizing and multiple pressure stages of expanders. [73]

A WHR-system that operates on the principles of Rankine and that is limited to the surrounding of an automobile needs to fulfill more restrictive boundary conditions than for example a Rankine cycle used in a power plant. First of there is a vast difference through the mobility aspect of the system as explained earlier. [73]
Therefore the addition of auxiliary heat exchangers like economizers and reheaters is very restricted, despite their positive influence on efficiency.

As the WHR system will most likely be needed to fulfill CO_2 regulations, it must be designed in a way to work for the lifespan of the car and needs to be reliable to prevent car failures caused by itself. This means that for a regular car the system must work for at least 1000 h without major overhauls. The system might be serviced regularly, however constant supervision as in power plants is not possible. [73]

Additionally there are thermodynamic boundary conditions restricting the system. The temperature level and mass flow of the exhaust gas of the engine are heavily dependent on the engines operating point, which changes all the time due to alterations in driving speed and road conditions. Thus the WHR-system needs to be able to adjust very quickly to these changes. The temperature level of the exhaust gas ahead of the EGX alters between 370-920°C (except for cold start) and the mass flow between 20-600 kg/h for a 2 l gasoline engine under stationary conditions. Thus an optimal system design for all operating points is impossible in the field and compromises have to be made. The common way is to design the system for medium engine loads that may occur during highway driving and to leave out high load points by using a by-pass for the heat exchanger. Thus the efficiency of the system in low-load operating points is acceptable and the system size/weight is confined. [73]

The condensation parameters are also different, if a Rankine system is integrated into a car. The condensation temperature is given by the temperature level of the heat sink, which in a car would be the ambient air. The temperature of the coolant after the radiator (warm engine) depends on the engine speed and the engine load and is typically somewhere between 40-80°C. This temperature level is much higher than that of heat sinks used in power plants, which are normally cooled by water from rivers or cooling towers. Higher condensation temperatures forbid the implementation of small low pressures as the risk of cavitation is increasing with temperature. Also the use of vacuum pumps to generate pressures below environment pressure are unlikely in mobile use thus low-pressures are somewhere between 0.5-2 bar (depending on the working fluid), whereas power plants use low pressures of 0.1 bar and lower [110]. [17], [73]

Even though the temperature levels of the heat source are not that much different than for example in a power plant, the heat flow of the source is much smaller due to lower gas mass flows (at nominal power: gasoline engine from a passenger car $\approx 6 \times 10^2$ kg/h $\Leftrightarrow$ 800 MW power plant: $\approx 2 \times 10^6$ kg/h) [101]. This means that the power output is much smaller, due to a smaller flow of working fluid that can be vaporized. The efficiency of such a downscaled Rankine system is lower because of increasing rates of heat losses and leakages, but also because of simpler system configurations and the demand for dynamic operation in mobile applications. [73]

Ambient conditions are another characteristic that needs to be considered when discussing the boundary conditions of mobile applications. While the cold start of a Rankine-system in power plants represents a very small portion of the operating states, it is a relevant state in passenger cars. Cold starts happen regularly and under very different conditions. Temperatures down to -30°C have to be considered for European use. Legislative requirements like real driving emissions have to be met for temperatures down to -7°C [24]. Furthermore temperatures down to -40°C must be withstood by the inactive system without any damage. This prohibits the use of pure water as working fluid, which is the standard fluid in power plants, as it would freeze below 0°C. [73], [42], [57]

2.2.4 Utilization of Feedback Power

The power that is regained through the waste heat recovery system can be fed back in more than one way. The first way is the direct use via a mechanical connection to the drive train. With this configuration there is no need for a generator, but the expander has to be placed in a way allowing for a connection to the drivetrain, for example next to the transmission or close to the engine belt. This limits the freedom in system design and has certain disadvantages considering the distance from the place of steam production (exhaust) and the placing of the expander which may result in additional pressure and heat losses as well as additional weight. The efficiency of this solution depends on the transmission efficiency from expander speed to crankshaft speed and the heat and pressure losses.

The next possibility is to use the power mechanically but to first convert it into electric energy via a generator and then use it with an electric motor. This gives additional freedom for the positioning of the expander, which may reduce the described heat and pressure losses but adds losses through the double conversion within the generator and the electric motor. The system that is used in this project is equipped with an expander that is directly coupled with a generator. The mechanical power is thus unknown but the electric power output is measured. To describe any advantages in fuel consumption for the engine through this possibility, some assumptions have to be made which will be explained in the following. The power that is produced from the generator comes as three-phase electric power. The frequency and voltage level of the electric power change through varieties of the expander speed. Thus an AC/DC converter is needed to transform the power into direct current for the electric motor. The efficiency of the used system was approximated to be 95 % by the designer. The efficiency of electric motors is said to be at around 95 % for most rotation speeds in common literature [90]. The possible reduction of needed engine torque can be calculated with equation 2.22.

$$\Delta M_{E-Motor} = \frac{P_{\text{WHR}} \cdot \eta_{AC-Converter} \cdot \eta_{Emotor}}{2 \cdot \pi \cdot n_{Engine}} \qquad (2.22)$$

The resulting torque demand on the engine can be calculated with the correlation:

$$M_{d,i} = M_d - \Delta M_i \qquad (2.23)$$

As there is no direct feedback through an electric motor on the setup of test bench, the resulting reduction in fuel consumption can only be simulated in GT-Suite with an engine model (as described in chapter 3.3) or by measuring the fuel consumption (on the test bench) with reduced torque demand. The result can be compared to the fuel consumption without feedback to give a first impression of the fuel savings. The reason for this is that with reduced torque demand the exhaust mass flow and temperature from the engine are also reduced, which results in lower power output from the WHR-system. This feedback loop is also described by Kupferschmid [60]. To increase the precision of the outcome more iterations of the tests/simulations have to be performed in order to reduce the changes in power output and fuel consumption. Co-simulation can avoid this problem, but also increases simulation time by magnitude and was thus omitted.

The third possibility to use the power from the system is to take load of the onboard DC-generator, which is considered by Horst [46] to be the most efficient way due to the lack of energy conversion steps. In a passenger car the average load on the generator is around 750 W [47]. This power is used for different electric consumers within the car like head lights, radio, electronic control unit (ECU), entertainment systems etc.. The generator in a regular car is driven by the engine belt and supplies the car with a $\approx$14 V power output. The efficiency of the generator is well dependent on the engine speed as well as the demanded load (normal operation values lie around 75 % [37]). The torque that the generator demands from the engine to produce a certain amount of electric power is given by equation 2.24.

$$\Delta M_{Generator} = \frac{P_{\text{Demand}}}{\eta_{Generator} \cdot 2 \cdot \pi \cdot n_{Engine}} \qquad (2.24)$$

With the calculated torque it is possible to measure/simulate the fuel consumption with the same electric power output produced by the generator and thus calculate any fuel savings through the WHR-system. This formula also visualizes the worth of electric energy in a car: any demand for it has to be delivered by the generator, which transforms mechanical power into electric power. Said mechanical power descends from the engine, which transformed it from chemical energy (fuel). This constitutes a long and loss-making energy conversion chain.

A last possibility to use the generated power is to store it. This could be done for example by charging the battery. As this does not directly influence the system efficiency and does more

3 Test Setup

To investigate the behavior of a WHR-system in steady-state and in cold start two approaches were followed. First an experimental setup was designed using an engine test bench including a real WHR-system and secondly a simulation model was created to perform tests that the real system would not be able to and to gain results for values that can not be measured on the test bench.

3.1 Test Bench

The experiments were performed on a fully automated test bench, that was equipped with an electric dyno (brake). The dyno was capable of dynamic operation for all possible engine loads and speeds. The test bench also had a dedicated cooling system installed, which was capable of cooling the test chamber down to -10°C.

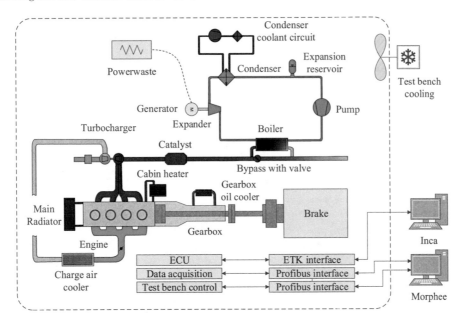

Figure 3.1: System schematic of the engine test bench with Rankine-system

The automation software used was *Morphee* from the company *d2t*. All automation routines, except for the basic ones like cooling and throttle, were created especially for this project. These included a routine for the regulation of mass flow, steam quality, cabin heater heat flow, expander load, condenser cooling, EGX bypass system and driving cycle. The test bench data was also recorded using the internal method of Morphee. The electronic control unit (ECU) of the car was connected to a laptop, where the communication and data acquisition was performed with *INCA*, a software from the company *ETAS*. Any changes to the standard engine application were implemented this way.

3.1.1 Engine

The engine was taken from an production car. It was mostly kept in the original state, the only changes that were made were the inclusion of measurement devices and changes to allow the operation on a test bench (see table 3.1). This engine was chosen due to its state-of-the-art technology and its use in mid-size passenger cars.

Table 3.1: Technical data of the engine

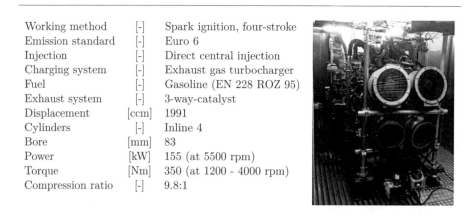

Working method	[-]	Spark ignition, four-stroke
Emission standard	[-]	Euro 6
Injection	[-]	Direct central injection
Charging system	[-]	Exhaust gas turbocharger
Fuel	[-]	Gasoline (EN 228 ROZ 95)
Exhaust system	[-]	3-way-catalyst
Displacement	[ccm]	1991
Cylinders	[-]	Inline 4
Bore	[mm]	83
Power	[kW]	155 (at 5500 rpm)
Torque	[Nm]	350 (at 1200 - 4000 rpm)
Compression ratio	[-]	9.8:1

Without a surrounding car, the engine lacked a cabin that would need to be heated and a cooling airstream. The cabin heat demand was recreated with a water cooled heat exchanger that took a predefined heat flow from the engine's coolant circuit. The heat flow setting was based on a conventional mid-size passenger car's simulation data[1] and was kept the same for each test run. The airstream was emulated with four fans in front of the engine (visible in the figure in table 3.1). The airstream was adjusted to fit the measurements of the air flow within the engine compartment of a real car at a driving speed of 50 km/h (further information can be found in the work of Röder [86]). Charge air and engine coolant were also conditioned with water cooled heat exchangers. The coolant and air passages were otherwise unaltered. The coolant was a mixture of water and Glysantin in a weight ratio of 1:1.

The transmission was performed with an automatic gearbox. The gearbox was fixed at the direct gear so that engine speed and brake speed would be the same. The clutch was actuated automatically and closed at engine speeds above 900 rpm. A water cooled gearbox oil cooler ensured safe temperature levels.

The exhaust system of the engine was taken from series production and was not altered except for the temperature and pressure measuring devices. The engine was equipped with a wastegate controlled exhaust gas turbocharger and a three-way catalytic converter. The exhaust manifold ahead of the turbocharger was double walled ex factory. Additionally the piping between turbocharger and catalyst and between catalyst and EGX was insulated. The exhaust system

---

[1]The simulation results were provided by a cooperation partner.

was also equipped with a bypass that could be used, when the remaining heat flow from the condenser to the coolant system was too high, because of a too high exhaust gas heat flow or if the WHR-system had any kind of failure.

3.1.2 Waste Heat Recovery System

The engine was equipped with a Rankine-WHR-system which consisted of components that were as close to production as possible. The focus of the design was set on the low to medium engine loads, which results in a maximal system power output of around 2 kW. By this approach it is possible to have an acceptable system efficiency at lower loads and during cold start, and a high efficiency at medium engine loads. The downside of this design focus are reduced maximally recoverable power outputs at high engine loads, which are only seldomly achieved. The components were chosen and designed in the previous work of Franke [27] and Hartmann [42]. The Rankine-system was positioned directly downstream of the catalyst to achieve the highest possible exhaust gas temperatures (see figure 3.2). A comparison of working principles of the expander, pump and working fluid was not the topic of this work and thus they were considered as given. The main point was to find influences of operating parameters and the interaction of engine and WHR-system.

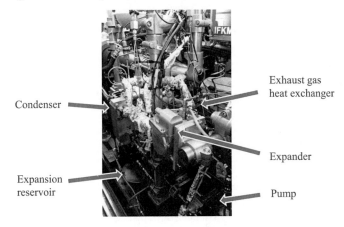

Figure 3.2: Working Rankine-system on the engine test bench

As the components of the system are not the focus of this project, they will only be briefly presented. As mentioned the WHR-system was built from components as close-to-production as possible. These components were readily purchasable with the exception of the EGX which was specially designed for the use in a Rankine-system. The other components present a cost efficient solution, but don't reach maximal thermodynamic efficiency.

The integrated pump was a gear type industrial pump. Exhaust gas heat exchanger and condenser were designed as plate heat exchangers. The expander was a scroll expander directly combined with a generator. A depiction and explanation of the working principle can be found in appendix A.1. As an expansion reservoir a 2 l membrane tank was used, which could be pressurized on its top. The gas inside was separated from the working fluid by a metal membrane. All the components were connected using stainless steel pipes that were isolated with glass wool where possible. The circuit was also equipped with two filters, a gas separator and could be vented at three points (highest point, expander and after the condenser). A schematic depiction of the WHR-system can be found in chapter 4.1

The operating point of the engine (torque, speed) and the set points of the WHR-system (steam quality, expander load, low-pressure) were kept constant by the test bench automation system for each test run. The power output of the WHR-system was measured, but was not used in any way and did not change the engine operation point. A power waste was constructed for this purpose. The alternating current from the generator part of the expander was first rectified and then smoothed. The load on the generator was controlled by clocking a water cooled resistor and thus generating different effective resistances over which the power was dissipated to heat. The working fluid used for the investigations was defined in earlier projects and was considered given [42]. It was a mixture of water and ethanol with an addition of 10 % silicon oil for lubrication. The mixture was around 3:1 water and ethanol (mass). Mixtures of fluids have zeotropic properties during vaporization. Their temperature and pressure are not constant while vaporizing but incline which can be considered beneficial in terms of exergy efficiency as explained by Chen [11]. The ethanol part was added to grant a certain protection against freezing which could potentially damage the system. Other substances were investigated within the chapters 4.1 and A.5. The results that are presented there, are meant to represent statements that are as general as possible, thus making the results applicable for other substances. Further details about the construction of the WHR-system, its components and the test bench can be found in the works of Schneider [89] and DellaBona [15].

3.1.3 Measuring Devices

The whole setup was equipped with a variety of measuring instruments. Temperatures were measured using PT100 resistance thermometers ahead and after each component to gain information on temperature changes. Based on this data and the working fluid mass flow, which was measured using a Coriolis scale, it was possible to calculate the heat flows of every part of the system. In order to determine the steam qualities and pressure losses, it was necessary to measure the pressure at each stage of the process. This was done with piezoresistive pressure sensors that detected the absolute pressure. The engine was also equipped with temperature sensors, pressure sensors and volume flow measurement turbines to determine heat flows and temperature curves for each component. The fuel consumption was recorded by a consumption measurement system. The engine torque was recorded with a torque transducer at the end of the gearbox. An analysis of the measurement uncertainty of the most important measuring devices can be found in appendix A.8.

3.2 Control Systems

In order to provide the demanded steam quality for all operating points and boundary conditions, a control system has to be implemented that controls the mass flow of fluid (and thus steam quality) in the Rankine-system automatically. A PID-controller works well in steady-state and slow transient operation, but is not able to reliably set the needed mass flows for transient conditions. The precision of the control system was of highest importance for the investigations in this project. Reaction time was only secondary, as the test cycles presented in this work do not include highly dynamic driving situations. Yet several transient cycles were tested that needed a fast control system. More information on control systems and strategies can be found in the works of Kupferschmid [60] and Liu [69].

Steam Quality: Feed Forward Control
The easiest kind of feed forward control, in order to reach a wanted setpoint for steam quality
is to measure the heat flow from the exhaust gas to the working fluid and to adjust the mass
flow of fluid according to the state (enthalpy) of the fluid ahead of the heat exchanger and the
desired state (enthalpy) after it:

$$\dot{m}_{\text{WF, Set}} = \frac{\dot{Q}_{EGX}}{h_{\text{WF},a,Ex} - h_{\text{WF},b,Ex}} \tag{3.1}$$

Equation 3.1 gives the needed mass flow to reach the wanted steam quality after the heat
exchanger for a given working fluid enthalpy ahead of the exchanger and a given heat flow. This
has to be considered only as an approximation, as the heat flow through the heat exchanger
depends on the mass flow, but for small changes in mass flow this approximation works fine, as
the heat flow does not change to quickly. With a fast recalculation of the mass flow after each
change this requirement can be met. Thus this kind of feed forward control works for controlling
the steam quality in car applications.

$$h_{\text{WF},a,Ex} = f(p, x_{Set}, \text{WF}) \tag{3.2}$$

$$h_{\text{WF},b,Ex} = f(p, T, \text{WF}) \tag{3.3}$$

The enthalpy of the working fluid before the heat exchanger depends on temperature and pressure and can be calculated through a working fluid database (like REFPROP). The wanted
enthalpy after the expander depends on the pressure after the exchanger and the requested
steam quality which is given through the setpoint. The pressure can be measured, but will
certainly change with an adapted mass flow. As long as the change is small or the recalculation
quick, this does not affect the control quality.
The precision of the measured heat flow is important for this kind of control. The lack of steam
quality deviation correction makes constant failures possible. To eliminate this possibility the
steam quality deviation is to be taken into account:

$$\Delta x = x_{Act} - x_{Theo} \tag{3.4}$$

$$x_{Set} = x_{Des} - \Delta x \tag{3.5}$$

The setpoint for the steam quality is calculated through the real externally desired setpoint
and the deviation value. This deviation value is calculated through the difference of the actual
measured steam quality (calculated with REFPROP) and the theoretical steam quality.

$$x_{Theo} = f(p, h_{Theo}, \text{WF}) \tag{3.6}$$

$$h_{Theo} = \frac{\dot{Q}_{Ex}}{\dot{m}_{\text{WF}}} + h_{\text{WF},b,Ex} \tag{3.7}$$

The theoretical steam quality is calculated from the current pressure after the exchanger and
the theoretical enthalpy of the working fluid after the heat exchanger that should result from
the current mass and heat flow. The internal setpoint for the steam quality x_{Set} is changed
according to the resulting difference of x_{Act} and x_{Theo}.
Failures through inaccurate heat flow measurement or other disturbances are now corrected. The
resulting control system is fast and precise and well suited for transient and constant operation
conditions in a passenger vehicle. Figure 3.3 shows the controlled steam quality in a driving
cycle. The steam quality deviation is kept below a value of 0.4 which translates to a temperature

deviation of 4.2 K at that point which was considered sufficient for the following less dynamic
tests as the fluid was not overheated nor subcooled.

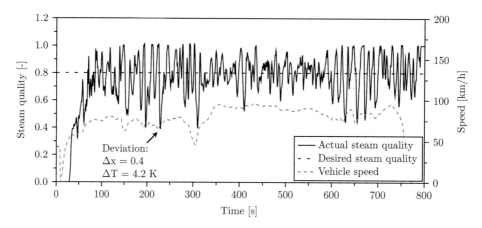

Figure 3.3: Behavior of desired and actual steam quality with the designed feed forward control in an
EPA *Highway Fuel Economy Test Cycle*

Mass Flow: PID

The mass flow that is calculated by the feed forward control for the steam quality is handed
over to a PID controller which generates a signal that is translated by a speed controller into
a rotational speed for the working fluid pump. For displacement pumps the mass flow is (almost) independent of discharge pressure and only dependent on rotation speed of the pump [57].
Thus the regulation can reach target values quickly. The resulting mass flow is compared to the
wanted mass flow and the rotational speed is changed accordingly.

Exhaust Gas Heat Flow: Security Functions

Security functions are needed for operating points outside the systems capabilities and for failures
within the system itself, to prevent damage to the system. These functions include increasing
low-pressure, decreasing condenser temperature, increasing mass flow and most importantly the
reduction of EGX heat flow.

The EGX heat flow can be reduced by rerouting the exhaust gas using a bypass. This cuts
steam generation almost immediately. Several conditions of the system were taken into account
that needed the bypass to be actuated: absolute exhaust gas heat flow, steam quality, working
fluid temperature after EGX and power consumption of the pump. If any of these conditions
were not within a defined limit, the EGX would be bypassed.

The timing of the deactivation of the bypass has to be handled with care, as the heat flow,
that is going to enter the system after that, is unknown and the exhaust gas heat flow can even
increase while the bypass was actuated if the engine operating point changes. The heat flow
entering through the EGX after the closing of the bypass valve is unknown, because of dynamic
of warm-up of the EGX and the deviation of current WF mass flow to the ideal WF mass flow.
A strategy was developed wherein the working fluid mass flow was increased to the maximum
while the bypass was open. As soon as none of the conditions were critical anymore the bypass
was closed. But the mass flow was kept at maximum for another 6 s, giving the system time
to measure the heat flow over the EGX. After this time the controllers were reactivated and
the mass flow was set again accordingly. This allowed for a smooth opening and closing of the
bypass without the knowledge of the current state of the engine and the exhaust gas heat flow.

3.3 Simulation

Simulations were performed to increase the data base of the experiments and to extend the drawn conclusions to other system configurations. *GT-Suite* Version 7.5 from *Gamma Technologies* was used for the combustion engine simulations as well as for the waste heat recovery system and the passenger cabin simulations. It is a CAE platform for multi-physics system simulations which is very common in the automotive industry and provides packages for thermalmanagement and waste heat recovery. It is a commonly used tool for the purpose of state calculation of WHR-systems and combustion engines [63], [78].

A detailed model of the WHR-system and a basic model of the engine and its exhaust system were created. The simulations were performed separately for both models. This prohibited interaction but allowed for much faster simulation speeds.

The models of the Rankine-system and of the exhaust gas heat exchanger will be discussed in the following. Further detail on the models of the combustion engine and of the passenger cabin can be found in appendix A.4.

Rankine-System Model

The Rankine-system model was designed according to the real setup of the system on the engine test bench. A two-phase approach is used in GT-Suite to simulate the Rankine-system which is a proven approach from the simulation of air conditioning systems [32]. The main inputs for the system are: mass flow of EG, temperature of EG, volume flow of condenser coolant and temperature of condenser coolant.

A components for the pump could be extracted from the library and was adjusted to the performance of their real counterparts. It was adjusted using the experimental data by Lagaly [61]. This included the mass flow and pressure increase in dependency of the rotational speed as well as the respective isentropic efficiency. The rotational speed of the pump was set by a PID controller within the simulation working in line with the strategy laid out in chapter 3.2.

The expander component was specially designed[2] to expand vaporous media. It was at first adjusted using maps for isentropic efficiency and volumetric efficiency given by the hardware manufacturer. These maps were most likely created based on computational fluid dynamic calculations and provided acceptable accuracy for the first simulations, but were adjusted by experimental data later on by Lagaly [61]. The rotational speed and torque on the shaft of the expander are set according to the mass flow and pressure of the steam entering the component.

The condenser was created as its own sub-model, consisting of a working fluid side and a coolant side. The plate heat exchanger design of the real component was applied to this design. Again the model was adjusted by experimental data [61]. The models for a preheater or recuperator configuration were designed similar to this model.

Expansion of the fluid through vaporization was accurately depicted by the simulation model and thus an expansion reservoir was also needed in the model. The component was dimensioned according to its real design. The pressure on the gas side was also set within the simulation model. A depiction of the Rankine-system model can be found in appendix A.4.

Exhaust Gas Heat Exchanger Model

The most relevant submodels for the overall precision of the simulation model is the model of the exhaust gas heat exchanger. This model is responsible for the amount of heat that is transferred from the exhaust gas to the working fluid and for the occurring pressure losses. The complex fluid guidance within the EGX demands a segmentation of the EGX into multiple heat exchangers. Through this approach it is possible to represent the different stages of countercurrent flow and

---

[2]by Gamma Technologies

cocurrent flow between exhaust gas and working fluid. The schematic of the model and of the
flow is depicted in figure 3.4. Further details about the construction and design of the EGX can
be found in the work of Bauer [5].

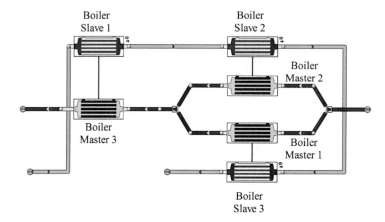

Figure 3.4: GT-Suite simulation model of the exhaust gas heat exchanger created for the system simulations. The WF flows 1-2-3 and the EG flows 1/2-3

The performance of each heat exchanger is mainly dictated by a characteristic map. This
map consists of mass flows, temperatures and pressures of both fluids at in- and outlet of
the EGX. GT-Suite uses this data to calculate heat flows and heat transfer coefficients for
Nusselt-correlations. The individual heat exchangers were calibrated with measurement data by
Lagaly. Further details can be found in his work [61].

Fluid Data Base

Most of the used fluid properties were provided by the GT-Suite database. The intake air for
the engine model and the engine and condenser coolant were given by that database and are
practically identical to the composition of the fluid used on the engine test bench. But GT-Suite
uses by default a surrogate gas for noncombustible air called "air2" which was used as the exhaust
gas that flows into the EGX. The specific heat capacity of this gas is the one of pure air thus
the value is too low for exhaust gas as exhaust gas also consists of considerable amounts of CO_2
and H_2O and has less O_2. Still air2 was kept as the used gas, because the model of the EGX
was validated with it.

The working fluid of the system was not included in the GT-Suite database and thus had to be
imported. The properties of the fluid were taken from a commercial fluid properties database
called *REFPROP* (Reference Fluid Thermodynamic and Transport Properties Database) from
the National Institute of Standards and Technology [66]. Due to a lack of detailed information
on the silicon oil, that was added to the WF ($\approx$ 10 % by volume) in the experiments, the oil
could not be taken into account for the simulations.

4 Results for Steady-State Operation

The measurements of the systems behavior in steady-state operation allow an in-depth look into the basic relationships between system parameters and configurations and how they influence the system. While many variables were measured and analyzed, the most interesting measured variable for these investigations was the electric power output of the Rankine-system which is supplied by the generator attached to the expander. This value will be referred to simply as the "power output" in the following. [73]

The engine operating points were kept constant for all measurements (except for chapter 4.3) and thus the efficiency of the system was directly linked to the electric energy that was recovered. The results will be presented for the two operating points from table 4.1 with the main focus on the 50 km/h operating point if not mentioned otherwise.

Table 4.1: Technical data of the engine

Settings	Operating point 50 km/h	Operating point 120 km/h
Theoretical car speed	50 km/h	120 km/h
Engine speed	1558 rpm	1843 rpm
Engine torque	37 Nm	161 Nm
Effective mean pressure	2.34 bar	10.16 bar
ECU	no intervention	no intervention
Coolant temperature	90°C	90°C
Charge air temperature	40°C	50°C
Cabin heating	no	no
Engine compartment air flow	no	no
Power feedback	no	no
Mean power-waste load	12.8 Ω	7.3 Ω
Condenser coolant flow	20 lpm	20 lpm

Power consumption of the pump is neglected for all results, as it does not change significantly for the variations and as its ratio to the power output of the system is generally around 10-15 %. The efficiency of the pump was measured to be somewhere between 5-20 % (depending on the operating point) which is not representative for the possible efficiency as the pump was designed with reliability in the focus and not efficiency. (> 70 %) [98], [99]. Thus the results are presented in gross amount. The power that is generated through the system is dissipated by the power-waste and thus does not influence the engine operation. [73]

This topic will be analyzed in chapters 4.3.7 and 5.1.6.

4.1 Influence of System Operating Parameters

The first tests were performed to analyze the influences of operating parameters of the system. These parameters are system variables that can be changed in a running system as explained in chapter 2.2.3. The influence of the three main operating parameters steam quality (mass flow), condenser temperature level and low-pressure will be discussed in the following chapter.

The tests were performed on the absolute basic configuration of the system which is illustrated in figure 4.1. The basic configuration consists only of the main parts: pump, EGX, expander and condenser. Not depicted are the expansion reservoir, the power-waste and all measurement devices. This configuration was chosen to limit any side effects and to gain a view into the basic relationships.

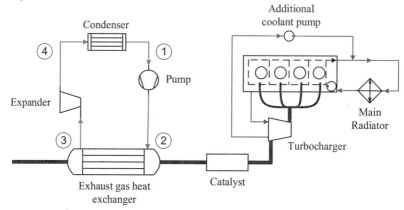

Figure 4.1: Schematic of the basic system consisting of the WHR-system, the exhaust gas system and the engine cooling system. The flow direction of the EGX is not depicted in the physically correct way (countercurrent) and the cooling circuit of the condenser is not represented, due to reasons of graphical representation.

4.1.1 Working Fluid Mass Flow / Steam Quality

Working fluid mass flow and steam quality are directly linked to each other for constant heat sources. The working fluid flows through the EGX, where it takes up the heat flow from the exhaust gas. In the progress of this it is heated up and vaporized depending on the amount of heat that is transmitted.

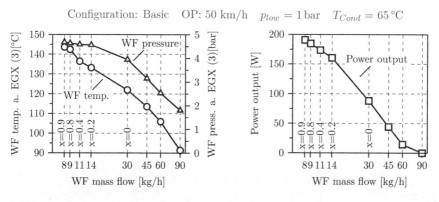

Figure 4.2: Influence of steam quality (mass flow) on temperature and pressure of the WF after the EGX (logarithmic x-axis base 10)

Figure 4.3: Influence of steam quality (mass flow) on power output of the system (logarithmic x-axis base 10)

As the heat flow does not change over different mass flow rates in the first approximation (5.03 kW for 9 kg/h to 5.31 kW for 90 kg/h), the steam quality can be directly controlled through the mass flow of working fluid. This behavior can be attributed to the design point of the EGX, which is designed for much larger heat flows. Thus it has overproportinally large heat transfer areas for low heat flows which it is thus able to transfer to very large fractions. Figure 4.2 and 4.3 show the influence of the mass flow on the WF pressure, the WF temperature and the power output.

As figure 4.2 shows, temperature and pressure of the working fluid, after the EGX, increase with decreasing mass flow, which results in higher steam qualities (depicted with their corresponding mass flow). The working fluid after the EGX reaches vapor state for mass flows up to 30 kg/h. After that the fluid remains liquid. But through the reduction in pressure within the expander the liquid can still be vaporized which accelerates the expander for mass flows up to 90 kg/h.

Even though the EGX heat flow is decreased for lower mass flows (through the higher WF temperature levels and thus smaller temperature difference to the EG) the power output is increased. This can be directly explained through the Carnot's theorem (equation 2.19). The higher the average temperature that the heat is added to the system (cf. [70]) and the higher the pressure ratio over the expander, the higher is the theoretical efficiency. The higher temperature level and greater pressure ratio overcompensate the loss in EGX efficiency. This behavior might change for higher steam qualities as the decrease in EGX efficiency rises, but this could not be tested as explained in the following.

The Carnot's theorem is not the only factor for the desired steam quality. The working principle of the expander and chosen kind of working fluid are also important. For scroll expanders, like the one used in the experiments of this work, the steam quality should be below 1. The wet steam helps the scroll in two ways: lubrication and sealing. Without a liquid phase both suffer and the scroll expander puts out less power and wears quicker (despite the added silicon oil), which is why this experiment was only performed for steam qualities up to 0.9. The steam quality for the remaining experiments was chosen to be 0.8 to give a certain security.

4.1.2 Low-Temperature Level

With the desired steam quality chosen in chapter 4.1.1 the system operating parameter "low-temperature level" was investigated. It was set to a certain value through conditioning the condenser coolant that enters the condenser. For adequate coolant volume flow rates and operating points way below the maximal design point, the estimation can be made that $T_{low} = T_{Cond}$ [72]. The desired value was set by a control system with a precision of $\pm$ 1 K.

Operating Point 50 km/h
Starting with the 50km/h operating point (OP) the condenser coolant temperature was varied from 35°C to 75°C in steps of 10 K. 35°C was the lowest temperature that could be set with sufficient precision and 75°C was the upper limit to prevent boiling at the condenser coolant exit and to stay below the condensation temperature of the WF for the low-pressure of 1 bar in order to avoid cavitation.

The depiction in figure 4.4 shows that the temperature ahead of the EGX is directly linked to the condenser coolant temperature. For temperatures below 55°C T_2 is higher than the condenser temperature due to the additional heat from the pump and from the surrounding exhaust system. For higher condenser temperatures the trend turns around and the working fluid looses heat to the environment. Isolation and environment conditions play a part in this, but for most configurations it can be said that under stationary conditions $T_2 \approx T_{\text{Condenser Coolant}}$.

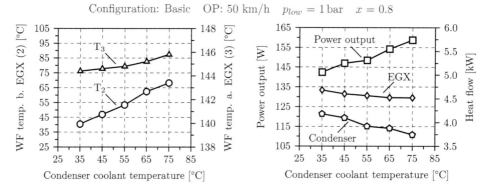

Figure 4.4: Influence of condenser coolant temperature on WF temperature before EGX (2) and after EGX (3)

Figure 4.5: Influence of condenser coolant temperature on power output as well as condenser and EGX heat flow

The temperature of the WF after the EGX is also proportional to the condenser coolant temperature, however in a less direct way. For a 40 K increase in condenser coolant inlet temperature the temperature of the WF ahead of the EGX T_3 is increased by 1.3 K. The increase in temperature after the EGX would mean an increase in steam quality but this is corrected through a higher mass flow, which results in higher pressure and thus constant steam quality.

Higher mass flow, higher temperature and increased pressure result in higher power output for higher condenser temperature levels. When the condenser coolant temperature level is increased the working fluid is less sub-cooled for a given low-pressure. That means less heat is taken from the system while the working fluid is still condensed. Due to the lower amount of sub-cooling and thus higher temperature when entering the EGX, less heat is needed to bring the fluid up to boiling temperature. Thus it is possible to reach higher temperatures after the EGX and higher mass flows. The EGX heat flow is reduced by the higher entrance temperatures of the WF, but the reduction of condenser heat flow overcompensates that loss, as shown in figure 4.5. Therefore the net heat input of the system is increased.

Operating Point 120 km/h

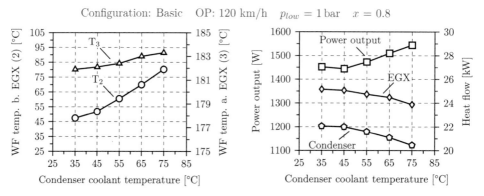

Figure 4.6: Influence of condenser coolant temperature on WF temperature before EGX (2) and after EGX (3)

Figure 4.7: Influence of condenser coolant temperature on power output as well as condenser and EGX heat flow

Figure 4.6 displays the same direct link between condenser coolant temperature and the temperatures of the working fluid ahead and after the EGX. The working fluid is not cooled to the same level as the entrance temperature of the condenser coolant as the condenser is almost at its maximum designed heat flow level. Thus the temperature ahead of the EGX is always higher than the condenser temperature. Still the approximation $T_2 \approx T_{\text{Condenser Coolant}}$ can be applied. The behavior of power output, EGX heat flow and condenser heat flow is similar to the 50 km/h OP as demonstrated in figure 4.7. The decrease in condenser heat flow overcompensates the reduced heat flow of the EGX which leads to higher power output. These results were also confirmed by simulations in the work of Lagaly [61].

System efficiency

The lowering of EGX efficiency can be seen in figure 4.8 for the 50 km/h OP (120 km/h OP shows a similar behavior). The EGX performance is deteriorated through the higher entrance temperature and thus lower temperature difference to the exhaust gas. This effect is not compensated through higher mass flows. But the system efficiency and the overall system efficiency are still increased with higher condenser coolant temperatures, as more heat is kept within the system while the heat source is constant.

In conclusion it can be said that the condenser temperature level should be as high as possible for a given low-pressure of the system. The limiting factor is the boiling point of the

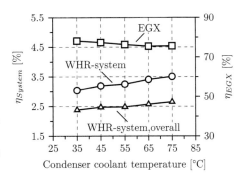

Figure 4.8: Influence of condenser coolant temperature on different system efficiencies for 50 km/h OP

coolant after the condenser. If the desired coolant temperature is not available, the coolant flow can be reduced to lower the cooling performance of the condenser and to increase the outlet temperature of the working fluid.

Simulation Results

Configuration: Basic   OP: 50 km/h $p_{low} = 1\,\text{bar}$ $x = 0.8$

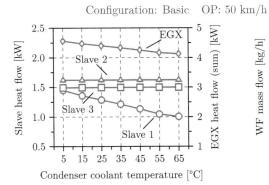

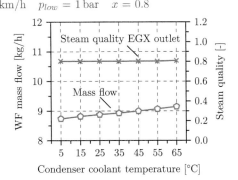

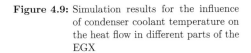

Figure 4.9: Simulation results for the influence of condenser coolant temperature on the heat flow in different parts of the EGX

Figure 4.10: Simulation results for the influence of condenser coolant temperature on working fluid mass flow

Figure 4.9 illustrates simulation results of the heat flows within different parts (slave 1-3) of the EGX and its overall heat flow. The position of the slaves (imagined segments of the EGX) can be taken from figure 3.4. The heat flows in slave 2 and 3 are almost unfazed by the change in the condenser coolant temperature level. The higher mass flow through constant steam quality seems to compensate the higher entrance temperature of the working fluid in these parts (see figure 4.10). Also the temperature difference between the working fluid and the exhaust gas is very high for these parts: $T_{EG,Slave2+3,in} = 450°C$, $T_{WF,Slave3,out} \approx 143°C$.

Slave 1 on the other hand shows a high dependence on the working fluid entrance temperature. The temperature of the exhaust gas entering it is much lower than that entering slave 2 and 3: $T_{EG,Slave1,in} \approx 175°C$. Thus the temperature difference compared to the working fluid is radically lowered when the condenser coolant temperature is changed from 5 to 65°C. The exit temperature of the WF after slave 1 is around $T_{WF,Slave1,out} \approx 133°C$ for this case.

While there is less exhaust gas heat transferred to the fluid, there is also less heat needed to preheat[1] the WF. Thus the enthalpy flow of the steam is increased through higher outlet temperatures and mass flows which overall increases the power output of the system. These results should be applicable to other heat exchanger designs as the temperature profile of the exhaust gas should not be too different for other countercurrent flow heat exchangers. Also because the difference in behavior seems to happen within the simple WF inlet part of the EGX (slave 1) - not the flow split part (slave 2+3).

The behavior of enhanced system power output through higher condenser coolant temperatures can not only be observed for the working fluid mixture of water and ethanol but also for other fluids. Similar simulations were performed for the working fluids pure ethanol (~isentropic WF) and R245fa (dry WF). All of them showed increasing power output with higher condenser temperatures, as can be observed in figure 4.11.

As the simulations were performed with the same model that was calibrated for the mixture of water and ethanol , the absolute values have to be considered with care, but the tendencies are evident. Also this is not a comparison of working fluids, as for example R245fa would most likely be much more performant in a cycle design with superheated steam and another kind of expander.

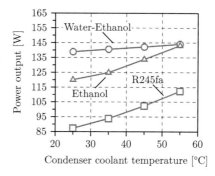

Figure 4.11: Simulation results for the influence of condenser coolant temperature on the power output of different working fluids. Low-pressure is 1.5 bar for water-ethanol and ethanol, and 6 bar for R245fa

4.1.3 Low-Pressure Level

The low-pressure of the system is the pressure level after the expander downstream to the pump (points 4-1). There are minor differences along the flow direction due to pressure losses and height differences, but the main level is determined by the air pressure on the expansion reservoir. The other side of the system (points 2-3) represents the high-pressure level of the system. The low-pressure was varied through connecting either a vacuum pump or an air compressor to the

---

[1]mostly done by slave 1

air side of the expansion reservoir. In this way it was possible to adjust the low-pressure from 0.5 bar to 1.5 bar absolute pressure. This was done manually with a precision of ± 10 mbar.

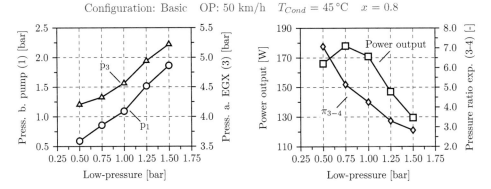

Figure 4.12: Influence of low-pressure on WF pressure before pump (1) and after EGX (3)

Figure 4.13: Influence of low-pressure on power output and pressure ratio over expander (3-4)

As mentioned earlier the pressure ahead of the pump (p_1) is directly in relation to the pressure on the gas side of the expansion reservoir (the low-pressure). Figure 4.12 shows the influence for a variation of low-pressure. The deviations between low-pressure and p_1 can be explained through additional hydrostatic pressure (the pump is located at the lowest level of the WHR-system) and additional back pressure from the piping. The overproportional increase in p_1 for low-pressures above 1 bar could not be explained entirely in this project. A connection to the elasticity of the membrane of the expansion reservoir is suspected to be the reason[2]. The same behavior can be observed for the 120 km/h OP. The influence on the pressure after the EGX (p_3) is almost as linear as the pressure ahead of the pump. There is only a divergence for low-pressures under 1 bar. The behavior for 120 km/h is similar again.

Even though both high-pressure levels (p_3 and p_2) and low-pressure levels (p_1 and p_4) within the system increase in a resembling way with increasing low-pressures, the pressure ratio between the two decreases. An example for the pressure ratio over the expander (π_{3-4}) is given in figure 4.13. The higher the low-pressure the lower the pressure ratio. The influence of pressure ratio on power output is in principle direct. The higher the pressure ratio the higher the potential expansion enthalpy difference (even though leakage rates increase). This can also be explained through the increase of working fluid temperature after the expander that can be observed with higher low-pressures [101]. The result is a smaller difference between high-temperature and low-temperature which according to the Carnot's theorem reduces efficiency (cf. chapter 2.1.5) [101]. Also for a set low-temperature the condenser heat flow increases with higher low-pressures (higher temperatures after the expander) which increases heat losses of the system.

Yet for a given expander there is an optimal pressure ratio dictated by its geometric expansion ratio as mentioned in chapter 2.2.2. This results in a loss of power output if the real pressure ratio differs from the optimal one. The outcome can be seen in figure 4.13. A clear peak in power output can be observed for a low-pressure of 0.75 bar at a pressure ratio of 5.1. This peak depends on the fixed geometric expansion ratio of the expander and the thermodynamic operating conditions. Expanders with different working principles or other optimal pressure ratios would have their maximal power output at a different low-pressure (and thus pressure

---

[2]Hysteresis or inconsistent resistance for different enlargements

ratio). If there are no means of altering the geometric expansion ratio of the expander by
design, the alteration of low-pressure can help adjust the pressure ratio to the desired ideal
value[3]. Summing-up it can be said that the low-pressure should be set as low as possible with
the restrictions of the optimal pressure ratio of the expander and the minimal condensation
temperature.

Figure 4.14 shows that the pressure at which the maximum power output for the 50 km/h
operating point occurs, is independent from the low-temperature in a reasonable range. It also
shows that higher low-temperatures are able to increase the power output at any low-pressure.
The same statement can be made for the 120 km/h OP, only that the maximum is shifted to
1.25 bar (see figure 4.15). At this point the pressure ratio is around 7.5, which is higher than
the optimal ratio.

The reason can be found in the significant change of the system operating point. The higher
EGX heat flow results in a higher WF mass flow to keep the steam quality constant. This
leads to a higher pressure (and temperature) ahead of the expander. If the low-pressure was the
same as for the 50 km/h OP, the pressure ratio would be even further away from the optimum.
Thus an increase in low-pressure improves the efficiency of the WHR-system. To reach the
optimal ratio the low pressure would have to be even higher. In this case 2.5 bar. But there is a
tradeoff between reaching the optimal pressure ratio and reaching low expansion temperatures
in order to achieve high Carnot efficiencies. At higher low-pressures the temperature level after
the expander increases which lowers the potential temperature difference over the expander and
also increases the heat losses over the condenser. The optimal low-pressure and pressure ratio
are thus dependent on the expander design and the current operating point of the system.

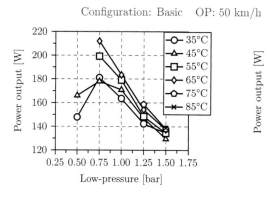

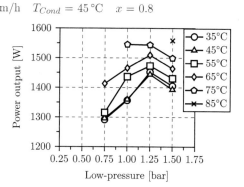

Figure 4.14: Influence of low-pressure and condenser coolant temperature on
power output for 50 km/h OP

Figure 4.15: Influence of low-pressure and condenser coolant temperature on
power output for 120 km/h OP

These results show, that the low-pressure of a system is one of its biggest levers for efficiency
improvement and that the low-pressure should not be set to a fixed value. The optimal low-
pressure depends on the operating point (heat supply) of the engine and the low-temperature of
the system for a given expander. Thus it should be variably set by a system that can calculate
the optimal setting under the given conditions with a certain security distance to the boiling
temperature: $p_{low} = f(T_{Boiling} + T_{Safety}, \dot{Q}_{EGX})$.

---

[3]which way is more effective/ efficient could not be investigated in this project

The influence of low-pressure on different system efficiencies is depicted in illustration 4.16. As the power output is the highest at 0.75 bar and the engine operating point is constantly at 50 km/h OP, the WHR-system has its maximum efficiency of 4.2 % at the same point. The overall system efficiency takes the EGX efficiency into account and reaches a level of 3.2 %. Due to the EGX efficiency being practically constant both curves are similar. Measurements proofed that these statements can be transferred to system configurations with recuperator and preheater.

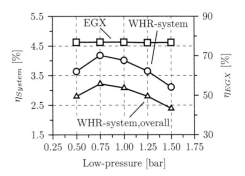

Figure 4.16: Influence of of low-pressure on different efficiencies for 50 km/h OP

A clear impact on EGX efficiency or mass flows could not be found. The efficiency is almost constant at a level of 77 %. It was expected that the measurements would show higher efficiencies with higher pressures, as the increase of pressure should have resulted in a shift in the position of the boiling point within the EGX. A possible explanation could be an unmeasurable decrease in mass flow to reach the same steam quality for higher low-pressures which equalized the efficiency increase. For the 120 km/h OP (peak at 1.25 bar) the basic behavior is the same, but a small increase in EGX efficiency was observed which could be attributed to the improved heat transfer due to the higher WF pressure within the EGX. Yet this improvement could not be seen in all measurements and thus bares uncertainty.

The simulation model was used to predict the influence of low-pressure on power ouput of systems with working fluids other than the water-ethanol mixture. The simulations were performed using the same models that were calibrated[4] for water-ethanol, only with the working fluid switch to pure ethanol and R245fa. Thus the results are not precise enough to compare the fluids, however an impact is visible. The simulation was performed with a condenser coolant temperature of 25°C. This allowed for direct comparison of the fluid despite their capital differences in boiling temperature.

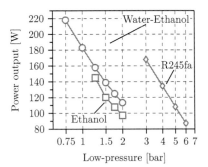

Figure 4.17: Simulation results for the influence of low-pressure on the power output of different working fluids. Condenser coolant temperature is 25°C (logarithmic x-axis base 10)

All fluids show the same basic impact of low-pressure on power output. The higher the pressure the lower the systems efficiency and thus the power output. For ethanol it was not possible to go below 1.25 bar without the simulation getting unstable due to the mass flow collapsing. For R245fa the low-pressure needed to be much higher (3-6 bar) to guarantee condensation.

---

[4]This calibartion included the control system parameters and the characteristic mappings for each component like pressure drop, Nusselt numbers and efficiency of the expander

4.1.4 Conclusion

The influences of system operating parameters were discussed within this chapter. These are parameters that are modifiable in a running system that have an impact on the system's power output. The focus was set on the main parameters: mass flow, low-temperature and low-pressure.

The mass flow directly controls the steam quality for a given engine operating point. Several settings were tested and the results showed that the steam quality should be as high as possible for the given expansion machine (in this case a scroll expander). Due to the needed lubrication of a scroll expander a steam quality of 0.8 was chosen as a compromise between power output and friction/wear.

Low-temperature level is given by the entrance temperature of the condenser coolant. It was found to be a key influence on power output. It should be as high as possible for a given low-pressure. Thus reducing heat losses over the condenser and thereby increasing the power output.

The low-pressure of the WHR-system can be controlled by setting the pressure level of the expansion reservoir and it should tendentially be as low as possible. However, the condenser heat losses and especially the geometric boundaries of the expander demanding a certain pressure ratio have an impact on it. Thus the low-pressure should be adjustable to the current boundary conditions of the system to improve its efficiency.

4.2 Influence of Additional Heat Sources

As mentioned before (chapter 2.2.3) the boundary conditions of a passenger car require that a WHR-system must be simple in order to fulfill the restrictions of packaging and weight. Yet it is still possible to add small heat exchangers provided they show potential in increasing the efficiency or the power output of the system and thus justifying to be added.

Four different configurations were chosen and created that used either additional heat from the system itself (partial flow and full flow recuperator), heat from the coolant circuit (preheater) or from the exhaust gas (turbocharger). Each configuration was tested on the test bench under stationary conditions to gain information on their potential. The configurations and the results will be presented in the following.

4.2.1 Partial Flow Recuperator

A recuperator is an internal heat exchanger within the system itself. The heat is transferred from the hot side of the circuit to the cold side. The hot source is the steam leaving the expander, that needs to be condensed anyway. The heat sink is the subcooled working fluid after the pump before the EGX. With this arrangement it is possible to have a working fluid temperature ahead of the pump that is below the boiling point and still reach a working fluid temperature after the pump that is higher than the condenser coolant temperature. This recirculation of heat is used in power plants to improve efficiency and power output of Rankine cycles and should be able to do the same in a mobile application. Further details can be found in the work of Franke [27].

The recuperator used for the first tests in this chapter is a partial flow recuperator that uses only a part of the hot steam. This way the required space is reduced and it is possible to combine the condenser and the recuperator into one unit. The unit is designed to provide save condensation of the steam and to have small or no impact on the pressure loss. The schematic of the partial flow recuperator integrated into the WHR-system on the test bench is presented in figure 4.18.

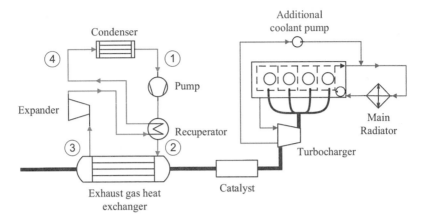

Figure 4.18: Schematic of the advanced system consisting of the WHR-system including a partial flow recuperator, the exhaust gas system and the engine cooling system

The detailed design of the condenser is presented in figure 4.19. Hot steam (light green) enters the condenser-recuperator unit through 14 flow passages. While flowing through them it is condensed. Most of the sheets are cooled by coolant (blue) from the opposing side in a countercurrent flow manner. Two passages are in contact with cold working fluid (yellow) which is almost at the same temperature level as the coolant entering the condenser (cf. chapter 4.1.2). The steam in these passages is also condensed, however its heat is not transferred to the coolant, but stays within the system. After the steam is condensed the working fluid is collected and is directed through one more passage for subcooling. After that it can be safely lead to the pump without any risk of cavitation or boiling.

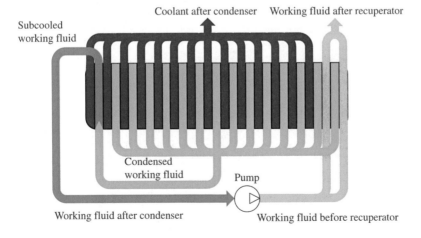

Figure 4.19: Layout and system integration of the partial flow recuperator

Influence on Temperature Levels

The heat transfer from hot steam towards the working fluid ahead of the EGX should help increase its temperature and thereby, according to the previous results, help enhancing the systems power output. To investigate the potential benefit from a partial flow recuperator a variation of condenser coolant temperature and a variation of low-pressure were performed. The results are compared to the previous results for a basic system without any additional heat sources from chapter 4.1.2 and 4.1.3.

OP: 50 km/h $x = 0.8$

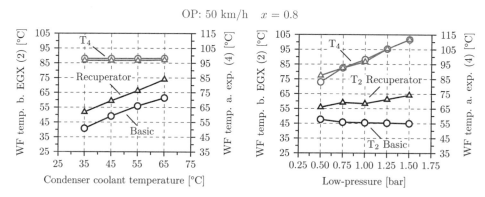

Figure 4.20: Influence of the addition of a partial flow recuperator on WF temperatures for different condenser coolant temperatures, $p_{low} = 1$ bar

Figure 4.21: Influence of the addition of a partial flow recuperator on WF temperatures for different low-pressures, $T_{Cond} = 45\,°C$

Figure 4.20 shows the behavior of the WF temperature after the expander and before the EGX for different condenser coolant temperatures. It proofs that the partial flow recuperator is indeed able to increase the temperature ahead of the EGX. The heat that is transferred by the recuperator is kept within the system and thus raises the working fluid temperature. The dependency on the condenser coolant temperature remains. The higher the coolant temperature the higher the working fluid temperature. The recuperator is able to add an offset to the basic temperature profile of around 10 K. Technically this offset should decline with increasing low-temperature level as the temperature difference between T_4 and T_2 shrinks due to the constant level of T_4. The measurement results for this range of low-temperature levels do not represent this behavior adequately.

The temperature level of T_4 is mostly dependent on the low-pressure of the system (as stated by Dietzel [17]). With inclining low-pressures as depicted in figure 4.21 the temperature after the expander is raised for both configurations. This increase in temperature difference between T_4 and T_2 is beneficial for the recuperator. The temperature before the EGX is enhanced with a recuperator while it is constant for the basic version. The temperature difference to the basic version increases from 8 K to 19 K for a low-pressure increase from 0.5 bar to 1.5 bar.

Influence on Power Output

The improvements in working fluid temperature ahead of the EGX through the use of the partial flow recuperator should amount to an increase in power output. The results for both the variation of condenser coolant temperature and low-pressure are presented in the following.

Figure 4.22 shows that the temperature increase through the regained heat by the partial flow recuperator results in enhanced power output of the WHR-system. The output is 4 - 10 W higher than without the recuperator which translates to an improvement of 2.2 - 5.8 % at this operating point.

OP: 50 km/h $x = 0.8$

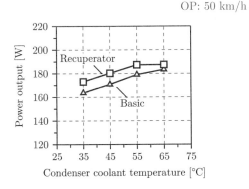
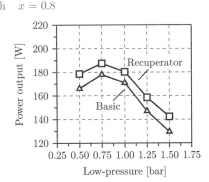

Figure 4.22: Influence of the addition of a partial flow recuperator on power output for different condenser coolant temperatures, $p_{low} = 1$ bar

Figure 4.23: Influence of the addition of a partial flow recuperator on power output for different low-pressures, $T_{Cond} = 45\,°C$

Despite the increasing thermal benefit from the recuperator at higher low-pressures the disadvantages of higher low-pressures cannot be compensated. The results are depicted in figure 4.23. The basic form of the power output curve is similar, but the recuperator is able to increase the output at every low-pressure by around 10 W. The highest output can still be gained at a low-pressure of 0.75 bar. The improvement at this point is 5.3 %.

The simulations display the same behavior. The partial flow recuperator is able to increase the temperature level ahead of the EGX by transferring heat from the low-pressure steam to the cold high-pressure fluid, which increases the systems power output. The recuperator does not include an additional heat source - still the only one is the exhaust gas which is constant in its values of mass flow and temperature. Thus the recuperator is not only able to increase the power output of the system but also its efficiency. Another benefit of the recuperator is that by reusing some of the heat of the remaining steam it is also possible to reduce the condenser heat flow and thus the load on the coolant circuit. Further details on the heat flows and simulation results will be presented in chapter 4.2.2 together with the results for the preheater. The tests could not be performed for the 120 km/h OP due to time limitations, yet there is no indication to suspect that the behavior would change. The statements should thus be transferable.

4.2.2 Preheater

Preheating (also called economizing) is a process that uses a heat exchanger that increases the temperature of the working fluid ahead of the main heat source. This reduces the effects of subcooling and brings the working fluid closer to its boiling temperature (in most cases without vaporization) [17]. Thus the heat of the main heat source, that is typically at the highest temperature level, can be used to a greater amount for vaporizing the fluid. In other words the exergy loss is reduced and the cycle is brought closer to a Carnot cycle [101]. This concept has already been suggested by Rankine himself [84]. The benefits are higher mass flows and/or higher temperature levels before the expander, which leads to an improved power output and, depending on the heat source of the preheater, can improve efficiency [20]. Simulations by Kim imply that benefits can be gained in mobile applications and that even more benefits can be found when using two preheaters [54]. Arias predicted that even larger benefits could be gathered if the engine was no longer supplied with coolant but directly with working fluid, thus increasing the preheating effect even further [1].

The tests in this chapter were performed with a single preheater ahead of the EGX that uses an electric coolant heater as its heat source. It is also possible to use the engine coolant from the engine outlet, but this was omitted for these investigations to avoid unwanted interactions. Using an electric heater is not reasonable in terms of the overall system efficiency, but it simplified the experiments. If the engine coolant is used, technically the efficiency of the WHR-system is not increased either, but the heat from the coolant can be considered free and thus the overall efficiency of the car is improved.

The coolant itself was circulated with a volume flow of 10 lpm and an inlet temperature of $T_{Pre} = 105°C$. With this approach the engine was still unaltered and the results of all configurations can be compared properly. The circuit of the preheater was designed in a way to make it possible to connect it to the coolant circuit of the engine for later tests (see 5.1.7). The heat exchanger of the preheater itself was a reused condenser heat exchanger that did not have the recuperator integrated. It was not specially dimensioned for this purpose but had sufficient heat transfer capabilities with a low pressure loss. A schematic representation of the preheater integrated into the WHR-system on the test bench is presented in figure 4.24.

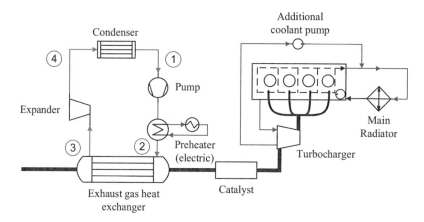

Figure 4.24: Schematic of the advanced system consisting of the WHR-system including an electric preheater, the exhaust gas system and the engine cooling system

A picture of the real preheater arrangement on the test bench can be found in figure 4.25. The volume flow and flow direction (countercurrent) were chosen to guarantee that the outlet temperature of the working fluid was practically identical to the inlet temperature of the coolant: $T_{Pre} \approx T_2$. With the addition of another heat source it is suspected that the power output of the system would incline but its efficiency would technically drop, as the heat input would be substantially increased and the suspected power output would not be increased in the same ratio. But when looking at the whole system, the coolant heat in a car at operating temperature can be considered "free". Thus the increase of WHR-system power output would again improve the overall efficiency.

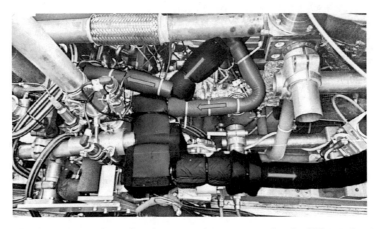

Figure 4.25: Implementation of a real preheater on the engine test bench: WF is isolated in red and heating fluid is isolated in black as well as the heat exchanger

Influence on Temperature Levels

Several tests were performed to investigate the influence of a preheater on a WHR-system. At first a variation of condenser coolant temperature was performed. The results are compared to the test with the partial flow recuperator and the basic system without additional heat source. The same comparison is drawn for a variation of low-pressure.

OP: 50 km/h $x = 0.8$ $T_{Preh} = 105\,°C$

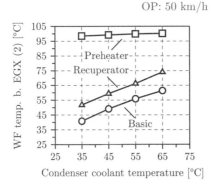

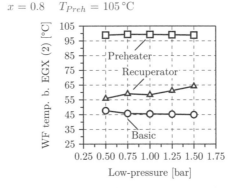

Figure 4.26: Influence of the addition of a pre-heater on WF temperatures for different condenser coolant temperatures, $p_{low} = 1\,\mathrm{bar}$

Figure 4.27: Influence of the addition of a pre-heater on WF temperatures for different low-pressures, $T_{Cond} = 45\,°C$

Due to the large amount of heat that is delivered, the working fluid before the EGX (2) can be heated to almost the temperature of the preheater itself as presented in figure 4.26. The temperature level is distinctively higher than the ones of the partial flow recuperator and the basic configuration. The dependency on condenser coolant temperature is lowered, with T_2 being almost constant. The magnitude of the dependency varies with the heat flow that can be supplied to the working fluid. For a regular engine cooling system at this operating point the heat supply can be considered sufficient. But the temperature level that can be supplied by a car might be lower. Tests with a preheater inlet temperature of $T_{Pre} = 95°C$ were also performed and showed similar results - only at a lower temperature level.

The results for different low-pressures are given in figure 4.27. The preheater shows no correlation on low-pressure just like the basic configuration. The dependency on T_4 that the recuperator owes its behavior to is not present. The temperature of the WF before the EGX reaches around 100°C for any low-pressure. This level is reached by the partial flow recuperator's heat source (T_4) only for low-pressures above 1.5 bar. Yet due to the partial flow design the heat source is not large enough to supply enough heat at that level which is why the partial flow recuperator still can't deliver WF temperatures at this level. This might be different with a full flow recuperator (chapter 4.2.3).

The results for the preheater are also valid for the 120 km/h operating point. The 10 lpm volume flow of heated coolant is a heat source large enough to keep the working fluid temperatures at the same level even though the mass flow of working fluid is six times higher (≈ 10 kg/h $\rightarrow \approx 60$ kg/h)

Influence on Heat Flow Levels
The preheater and also the recuperator influence the heat flows of the system. Higher temperatures ahead of the EGX lead to lowered temperature differences and thus lower heat flows of the EGX. And while the recuperator reduces the heat flow to the condenser by reusing the heat, the preheater increases the overall heat input and thus the heat flow over the condenser. The influences are depicted in figures 4.28 and 4.29.

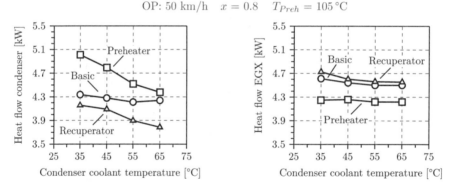

Figure 4.28: Influence of the addition of a preheater on condenser heat flow for different condenser coolant temperatures, $p_{low} = 1$ bar

Figure 4.29: Influence of the addition of a preheater on EGX heat flow for different condenser coolant temperatures, $T_{Cond} = 45$ °C

Basically the heat flow over the condenser is lowered with higher settings of the condenser coolant temperature. Less heat is withdrawn from the system and more heat can be reused to generate power. For the basic configuration the heat flow is reduced by 100 W from 35°C to 65°C. The heat input from the preheater is directly proportional to the condenser coolant temperature. The higher the low-temperature level the less heat can be put into the system by the preheater due to its constant temperature (105°C). Thus even though the mass flow and the temperature level of the working fluid is the highest with a condenser coolant temperature of 65°C, the condenser heat flow is the lowest (reduced by 630 W from 35°C to 65°C). Yet the preheater shows the highest amount of all configurations. The partial flow recuperator reduces the heat flow for all temperatures compared to the basic configuration. The condenser load is reduced by 370 W from 35°C to 65°C.

Due to the additional heat input, the preheater configuration is the only configuration with a higher condenser heat flow than EGX heat flow. The high working fluid temperature before the EGX reduces the temperature difference and thus lowers the heat flow. Nevertheless, the overall heat input is the highest. As the temperature level of the preheater is constant, its EGX heat flow is also. Partial flow recuperator and basic configuration show a dependency on the condenser coolant temperature. The reason can be found in the relation that higher condenser coolant temperatures lead to higher working fluid temperatures before the EGX and thus reduced heat flows.

Even though the mass flows are increased with higher temperatures before the EGX (to keep the steam quality constant), the EGX heat flow of the recuperator should still be lower than the one of the basic version due to the higher temperature level of the recuperator (T_2). A possible explanation might be given by the lubrication oil: a higher amount of oil leads to a higher mass and heat flow but might not reduce the power output noticeably. As the system had to be refilled with WF and oil for each configuration, it is possible that the ratio of oil and working fluid deviated from the target value. The following simulation results paint a clearer picture of the EGX heat flows.

Influence on Power Output

As explained in chapter 4.1.2: higher working fluid temperatures ahead of the EGX lead to higher temperatures and pressures after the EGX and also higher mass flows. This should result in a gain of power output through the addition of the preheater.

$$\text{OP: } 50 \text{ km/h} \quad x = 0.8 \quad T_{Preh} = 105\,°C$$

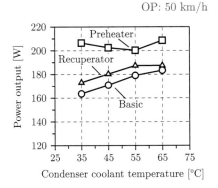 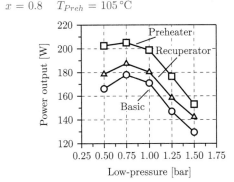

Figure 4.30: Influence of the addition of a pre-heater on power output for different condenser coolant temperatures, $p_{low} = 1$ bar

Figure 4.31: Influence of the addition of a pre-heater on power output for different low-pressures, $T_{Cond} = 45\,°C$

Figure 4.30 depicts the increase in power output of the WHR-system that is caused by the temperature improvement resulting from the preheater. The output is distinctively higher than the ones of partial flow recuperator and basic configuration. The difference is similar to the difference in working fluid temperature before the EGX (2). At the optimum point ($T_{Cond} = 65°C$) the output is increased from 183 W to 208 W which is an improvment of 13.7 %. The output should be almost independent from the condenser coolant temperature with a small enhancement at higher temperatures. The visible deviations can be explained through the complicated control of preheater temperature that resulted in unsteady conditions for the WF ahead of the expander. A variation with a preheater temperature of 95°C showed similar results. The power output was smaller than with 105°C but still higher than the one of the partial flow recuperator.

The output can also be improved for a variation of low-pressure. This can be found in figure 4.31. The preheater surpasses the partial flow recuperator and the basic version in all points. The best point for all configurations is unfazed by the additional heat input and can still be found at 0.75 bar at this operating point. The dependency on low-pressure remains.

Simulation Results

Simulations were performed for all three configurations. The most relevant results for EGX heat flow and power output are presented in the following.

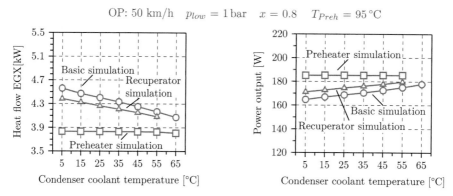

OP: 50 km/h $p_{low} = 1 \, \text{bar}$ $x = 0.8$ $T_{Preh} = 95 \, °C$

Figure 4.32: Simulation results for the influence of the addition of a preheater on EGX heat flow for different condenser coolant temperatures

Figure 4.33: Simulation results for the influence of the addition of a preheater on power output for different condenser coolant temperatures

As mentioned in the explanation of figure 4.29, the measurements could not represent the EGX heat flow of the partial flow recuperator correctly, but the simulation results in figure 4.32 do. The heat flow of the recuperator is lower than the one of the basic version due to the additional heat input and the higher temperature level ahead of the EGX. The simulations also show that the trend of higher heat flows for recuperator and basic would continue for condenser coolant temperatures down to 5°C and that the EGX heat flow of the preheater configuration would still be constant. Absolute values differ from the measurement results as the simulation does not calculate heat losses to environment while the measurements take these into account.

The simulated power output is depicted in figure 4.33. The results match the ones from the measurements in figure 4.30. The values are a bit closer together as the simulation of the preheater was performed with a setpoint of 95°C. The statements from the measurements are clearly confirmed. Additional heat input leads to higher temperatures before the EGX which lead to higher temperatures and pressures after the EGX. With a controlled setpoint for the steam quality this leads to higher mass flows. In the end this results in higher power outputs of the WHR-system.

4.2.3 Full Flow Recuperator

The results for the partial flow recuperator from chapter 4.2.1 show that using the heat of the remaining steam after the expander to increase the working fluid temperature ahead of the EGX is beneficial for the system's power output and also to reduce the condenser heat flow. However, the investigations with the partial flow lack an assertion about the maximal amount of heat that can be withdrawn, if the full flow is used. For this purpose a full flow recuperator was chosen and adapted to the system. This different design approach might lead to additional back pressure, space demand and weight and might increase the risk of cavitation due to the slower drop of temperature of the remaining steam, but might increase the power output even further.

In order to investigate the potential, a full flow recuperator was integrated into the system. This was achieved by including a plate heat exchanger between the condenser and the expander. The heat exchanger was around 200x80x40 mm (LxWxH) in size. The same variations of condenser coolant temperature and low-pressure were performed and the results are compared to the basic configuration and the preheater in the following.

Influence of Low-Temperature

The previous investigations proofed that the condenser coolant temperature has a direct impact on the temperature of the working fluid before the EGX for a system configuration that includes a partial flow recuperator or for a basic configuration. A preheater can be considered practically independent from it, due to the magnitude of its heat source.

$$\text{OP: } 50 \text{ km/h} \quad p_{low} = 1 \text{ bar} \quad x = 0.8$$

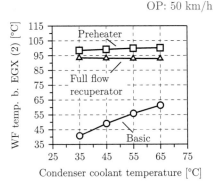
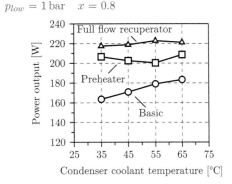

Figure 4.34: Influence of the addition of a full flow recuperator on WF temperature for different condenser coolant temperatures

Figure 4.35: Influence of the addition of a full flow recuperator on power output for different condenser coolant temperatures

Figure 4.34 shows that the full flow recuperator is able to distinctively enhance the working fluid temperature before the EGX and almost reaches the level of a 105°C-preheater, which is clearly higher than the one of a partial flow recuperator. The heat source of the full flow recuperator is not at a higher temperature level than the partial flow recuperator ($T_4 \approx 98°C = $ const.), but uses the full mass flow of steam. The enthalpy of vaporization that is "regained" through the condensation seems to be sufficient to keep the working fluid at a constant level of $T_2 \approx 93°C$. A dependency on the condenser coolant temperature cannot be found.

The power output of the WHR-system with a full flow recuperator is also constant for different condenser coolant temperatures, as can be seen in figure 4.35. Judging from its temperature level ahead of the EGX its power output should be just below the one of the preheater system. Yet it is clearly above it, with a power output of around 220 W which is the highest measured output for this OP. The reason for this disorder is an overhaul of the expander due to a bearing failure just before the experiments. This overhauled expander had reduced friction losses and thus a higher power output. The experiments with the full flow expander were the last ones performed, thus all other experiments were performed with the "old" expander with higher friction.

A point that is not represented by the 50 km/h results is the additional steam volume and system volume that is created by the full flow recuperator. While the partial flow recuperator replaced some of the condenser coolant channels, the full flow recuperator adds further channels and thus volume to the system. As the full mass flow of steam is used to heat up the working fluid, the condensation potential or, in other words, the potential to quickly withdraw heat from the steam is reduced. This is caused by the lowered temperature difference between coolant (in this case working fluid) and steam. The result is a longer time span that the steam remains at least partially vaporous. The increased amount of steam within the system demands larger expansion reservoirs which is disadvantageous in regards to packaging and weight. This also increases the distance that the volume flow of steam, which is much higher than the volume flow of fluid, has to travel. All in all this increases the pressure loss within the system and reduces power output. For the 50 km/h OP this is not noticeable as the geometric boundaries are overbuilt compared to the mass/volume flow. The pressure losses over the condenser are in the range of $\approx$ 10 mbar, but for the 120 km/h OP these losses are distinct. For example: at a low-pressure of 1.25 bar and a condenser coolant temperature of 45°C the basic system and the partial flow recuperator produce a pressure loss of 80 mbar. The full flow recuperator raises this value to 400 mbar. Due to the overhaul of the expander it was not possible to compare the influence on power output, but the impact should be noticeable. Creating a bigger full flow recuperator to reduce the pressure losses would increase the space demand and weight of the system which might reduce benefits. An optimized condenser, fit to the reduced condenser heat flow of the full flow recuperator, might reduce the pressure loss, but cannot compensate it completely.

Influence of Low-Pressure

As the temperature level after the expander is enhanced by higher low-pressures, it might be possible to improve temperature and power output of the system even further with a full flow recuperator at elevated low-pressures. A variation of low-pressure was performed to compare the results to the other configurations. The impact is displayed in the figures 4.36 and 4.37.

OP: 50 km/h $T_{Cond} = 45\,°C$ $x = 0.8$

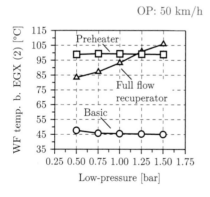
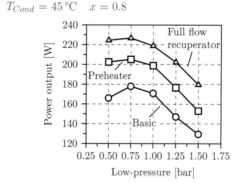

Figure 4.36: Influence of the addition of a full flow recuperator on WF temperature for different low-pressures

Figure 4.37: Influence of the addition of a full flow recuperator on power output for different low-pressures

The full flow recuperator shows a dependency on low-pressure in its temperature increase before the EGX much like the partial flow recuperator (cf. figure 4.21). For the full flow recuperator this correlation is much more direct than for the partial flow recuperator. The temperature before the EGX (T_2) climbs from 83.4°C to 105.9°C from a low-pressure of 0.5 bar to 1.5 bar (and a temperature after the expander (T_4) of 87.5°C to 111.0°C). This is an increase of 22.5 K for the full flow compared to the 8.4 K increase of the partial flow recuperator. At the highest low-pressures the full flow recuperator is even able to surpass the temperature improvement of the 105°C-preheater.

The curve of the power output over low-pressure for the full flow recuperator is similar to the preheater and all other configurations. When the low-pressure veers away from the optimum of 0.75 bar (for this OP) the power output is reduced. The improved temperature level through higher low-pressures does not compensate the disadvantages of them. Again, the comparison of the power output level to the other configurations is complicated due to the overhauled expander these measurements were performed with. At least the three measurement points for lower low-pressures of the full flow recuperator system should be located below the ones of the preheater system due to their lower temperature before the EGX.

The basic statements about independence of condenser coolant temperature, distinct temperature increase before the EGX and increase power output are also valid for the 120 km/h OP as further measurements indicate.

4.2.4 Turbocharger

The turbocharger of a combustion engine spins with multiple thousands of revolutions per minute and therefore normally has to be lubricated by oil. The high temperature levels within the turbine caused by the hot exhaust gas require some kind of cooling for the oil and the components of the turbocharger [35]. For this reason turbochargers in modern engines are often designed to feature a coolant connection. This coolant supply keeps the oil temperature within certain borders to secure its lubricity and to keep it from thermal disintegration. The heat is dispatched constantly and ranges somewhere between a few hundred watts and a few kilowatts depending on the engine operating point. An additional coolant pump is needed in most coolant circuit configurations for times when the engine stops and no coolant is supplied by the main coolant pump. This pump is added because of the heat remaining in the structure of the turbocharger: Even when the engine is turned off the oil still needs to be cooled and boiling of the coolant within the turbocharger must be prevented.

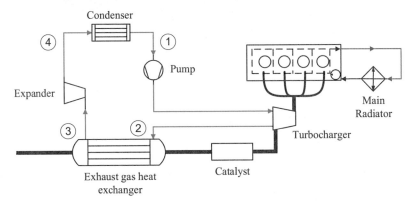

Figure 4.38: Schematic of the advanced system consisting of the WHR-system including the turbocharger as an additional heat source, the exhaust gas system and the engine cooling system

During the test series for this project the idea of using the turbocharger waste heat as an additional heat source for the WHR-system was developed and applied for patent[5]. The

[5]Patent pending; Title: „Verbrennungskraftmaschine mit fluidisch gekühltem Abgasturbolader und Abgaswärmetauscher"; Official Reference Number 10 2017 101 288.5

main idea was to directly use the turbocharger's coolant channels to transport working fluid through the turbocharger and thus heating it up. By choosing this design it is possible to gain the maximal amount of heat from the turbocharger without the losses of an additional coolant-to-working-fluid heat exchanger. It is also possible to drop the additional coolant pump by transferring the task of engine-off-cooling to the WF-pump. The coolant circuit of the engine is partially relieved from turbocharger heat, but as part of the heat will be transferred back by the condenser, this benefit can be considered small. A schematic of the suggested system design is presented in figure 4.38.

Some safety tests and preparations were performed beforehand. The coolant inlet ports of the turbocharger were modified to be able to connect them to the WHR-system. The turbocharger was cleaned to avoid pollution of the working fluid with remaining coolant. And lastly a pressure and leakage test was performed. This was achieved by blocking the outlet of the turbocharger and pumping water into it. Pressures up to 20 bar were tested and held for 5 min without any noticeable leakage. The thick-walled cast iron walls of the coolant channels are sufficient for the application as a working fluid heat exchanger. The turbocharger was fitted with additional temperature sensors and the coolant connections of the engine were directly joined together. A picture of the adapted and integrated turbocharger is given in figure 4.39.

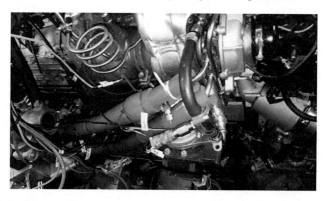

Figure 4.39: Turbocharger as heat source: connections to WHR-system are isolated in red material, previous coolant circuit is shorted out

Influence on the WHR-System

The turbocharger in this system design should basically behave like a preheater with a smaller heat source. It is an external heat source at a temperature level below the boiling point of the pressurized working fluid. The coolant volume flow of the turbocharger is normally around 1.2 lpm for the 50 km/h OP. The coolant temperature is raised by $\approx$ 5 K which amounts to a heat flow of $\approx$ 300 W. Several tests were performed to investigate the possibilities of the use of the turbocharger waste heat. A comparison to the preheater and the basic system at different condenser coolant temperatures is given in the following.

Figure 4.40 depicts the temperature curves of the working fluid before the EGX. By routing the fluid through the coolant connections of the turbocharger and thereby using its waste heat it is possible to distinctively increase its temperature. The temperature level is higher than the one of the partial flow recuperator, but not as high as the one of the full flow recuperator or the preheater. The higher the condenser coolant temperature the higher the working fluid temperature after the turbocharger. Yet the temperature increase is diminished for higher coolant temperatures. The reason is the temperature difference within the turbocharger is being lowered. As suspected the turbocharger does behave like a preheater with a smaller, lower temperature heat source. A dependency to the low-pressure could not be found; the turbocharger behaves like the preheater in this point. Also no influence on the exhaust gas temperature at the outlet of the turbocharger could be found. Thus the exhaust gas aftertreatment system is not influenced at all. The heat is dispatched anyway no matter whether it is gathered by the coolant or by the working fluid and due to the similar temperature levels of working fluid and coolant it is not possible to detect any impact.

$$\text{OP: } 50 \text{ km/h} \quad p_{low} = 1 \text{ bar} \quad x = 0.8$$

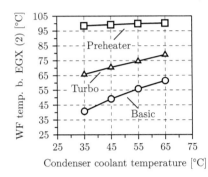 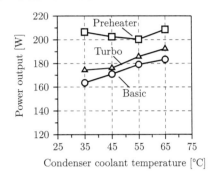

Figure 4.40: Influence of the addition of the turbocharger waste heat on WF temperatures for different condenser coolant temperatures

Figure 4.41: Influence of the addition of the turbocharger waste heat on power output for different condenser coolant temperatures

The resulting power output is presented in figure 4.41. The use of the turbocharger waste heat improves the system's power output. The improvement is not as large as with a full flow recuperator or a preheater, but around the level of a partial flow recuperator. Considering its T_2-level it should be higher than the one of the partial flow recuperator. This can be attributed to the wear of the expander, as the experiments took place almost at the end of the project. If the additional heat is taken into account, the efficiency of the system is technically reduced. But similar to the preheater configuration, the turbocharger waste heat can be considered free. In this case the efficiency of the whole system is improved.

Influence on the Turbocharger

The use of the turbocharger heat seems beneficial in all ways. No negative aspects for the WHR-system or the engine could be found for the steady state operation at the 50 km/h OP. Tests at higher car speeds and thus engine loads and speeds were conducted to find if this is also true for operating points with higher turbocharger heat flows.

The temperature of the working fluid at the outlet of the turbocharger (= temperature ahead of the EGX) was chosen as a criterion for the ability of the WHR-system to sufficiently cool the turbocharger. If the increasing heat flow caused by the higher engine load could not be rejected in a secure way, the temperature level of the working fluid at the outlet would increase and possibly reach unsafe levels. Any temperature of the working working fluid above 105°C was considered as too hot, as cooling by engine coolant would not reach higher levels. As figure 4.42 shows, the temperature is stable and almost constant for vehicle speeds up to 120 km/h. The heat dissipation and the temperature of the turbocharger can thus be considered stable as well. The direct integration of the turbocharger into the WHR-system can be regarded as save.

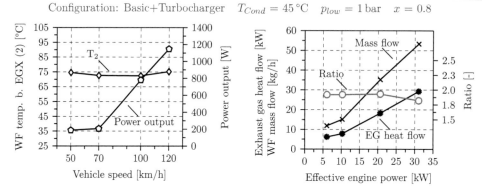

Figure 4.42: Influence of the addition of the turbocharger waste heat on WF temperature and power output for different driving speeds (Operating points)

Figure 4.43: Working fluid mass flow and exhaust gas heat flow for different effective engine power outputs

The power output of the WHR-system inclines with higher vehicle speeds due to the increased exhaust gas heat flow. The higher heat flow that is gathered by the EGX is answered by the system with a higher mass flow of working fluid in order to keep the steam quality constant. The behavior of both exhaust gas heat flow and working fluid mass flow is depicted in figure 4.43. Also given is the ratio between the two, which is effectively constant. This indicates that the cooling of the turbocharger is ensured by the steam quality control of the WHR-system. Thus is can be assumed that it is possible to run the system including a turbocharger at speeds above 120 km/h. If the system changes to bypass operation at higher exhaust gas heat flows, the pump has to keep running to ensure the cooling of the turbocharger. But considering the reduced pressure demand at this operation mode the additional power consumption can be neglected. However, the pump would most likely have to keep running anyways in order to be ready if the bypass operation is stopped and the WHR-system has to run again.

The direct use of working fluid to gather the turbocharger waste heat is advantageous to the working fluid temperature before the EGX resulting in increased power output. This improvement comes without disadvantages in steady-state operation. At the same time the system complexity can be lowered (or at least be kept) due to the possible omission of the additional coolant pump.

4.2.5 Conclusion

The results for different system configurations presented in this chapter show that it is possible to increase the systems power output by a definite amount using additional heat sources. Investigations were performed for partial and full flow recuperators, a preheater with a 105°C and 95°C heat source and a configuration that uses the waste heat of the turbocharger to its benefit. They all increased the working fluid temperature before the exhaust gas heat exchanger and thus improved the systems power output. They also had impacts on the heat flows of EGX and condenser. The exemplary depiction in figure 4.44 will be used to give an overview of their performance.

The basic system without additional heat source is able to achieve a heat flow of 4.50 kW over its EGX. 4.24 kW of that heat flow is rejected by the condenser after the expansion process. A total power output of 171 W can be gained with the basic configuration. The remaining heat flow - the difference between the input heat flow(s) and the output heat flows / powers - can be attributed to heat dissipation across the walls of the piping and components.

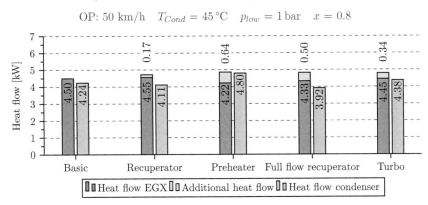

Figure 4.44: Comparison of the heat flows for different WHR-system configurations

The partial flow recuperator adds another 170 W to the heat input of the EGX heat flow (which should be lower than the one of the basic cf. 4.2.2). This increases the overall heat input to 4.72 kW, resulting in a power output of 180 W. At the same time the partial flow recuperator reduces the condenser heat flow to 4.11 kW owing to its internal heat transfer.

The preheater featuring a 105°C heat source produces the highest overall heat input, namely 4.86 kW, but has the lowest EGX heat flow of 4.22 kW. The additional heat input of 640 W results in the highest condenser heat flow of 4.80 kW. With the preheater having the highest overall heat input it is able to generate the highest power output of 203 W.

The full flow recuperator is almost able to reach the same overall heat input as the preheater. With 4.83 kW it is just below the value of the preheater configuration, which should result in a similar but lower power output. The advantage of the full flow recuperator lies in its internal heat transfer which reduces the condenser heat flow to the lowest overall value of 3.92 kW.

The turbocharger configuration reaches a heat input of 4.79 kW which is located between the partial and the full flow recuperator. Such is its condenser heat flow and power output[6].

These values should give an impression of the magnitude of the impact of the different configurations in terms of energy flows. The results are similar when looking at different system parameters and operating points. A comparison of the resulting system efficiencies of the different additional heat sources for the 50 km/h and 120 km/h operating points is given in figures 4.45 and 4.46.

The addition of heat sources to the system has distinct benefits for the system in terms of overall efficiency. Figure 4.45 and 4.46 display this impact for the 50 km/h and 120 km/h OP. The optimal settings for condenser coolant temperature and low-pressure[7] were chosen for this comparison. This way the maximal increase in system efficiency for each configuration and operating point can be presented. As the engine operating point was kept the same and the

---

[6]not at this specific point but in general
[7]no measurements for 120 km/h at 1.25 bar and 65°C were recorded for all configurations, thus the value for 1 bar is represented. 1.25 bar is considered to reach even higher levels

engine was not altered in any way, every increase in power output means an increase in the efficiency of the overall system "car".

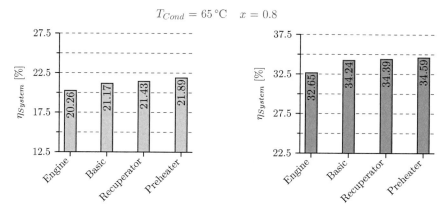

Figure 4.45: Comparison of system efficiency for different WHR-system configurations for 50 km/h OP at 0.75 bar low-pressure

Figure 4.46: Comparison of system efficiency with different WHR-system configurations for 120 km/h OP at 1 bar low-pressure

The engine itself provides energy at an efficiency of 20.26 % at this OP, which is quite low due to the high amount of charge exchange losses at such a low load. The WHR-system in its basic configuration is able to increase that efficiency up to 21.17 %. The addition of a partial flow recuperator adds another 0.26 %-points of efficiency to the system. Adding a preheater to the basic system enhances the output by 0.72 %-points.

At the 120 km/h OP the increase in efficiency of the basic system is higher than at the 50 km/h OP. This can be achieved despite the increased engine efficiency of 32.65 % at this higher load. The system is closer to its designed operating point and benefits from the higher mass flows etc.. The basic system enhances the system efficiency up to 34.24 %. The partial flow recuperator adds 0.15 %-points to that value and the preheater adds 0.35 %-points. The benefit of the additional heat sources at this higher engine load is reduced. The exhaust gas delivers a higher heat flow and the exhaust gas heat exchanger is not able to even use that heat completely to heat up its working fluid and produce higher steam mass flows. Thus additional heat can not help to reduce the exergy losses in the same manner as at lower engine loads. Yet they still prove to be advantageous.

A topic that could not be covered within these experiments is the use of an alternative working fluid. With the additional low-temperature heat source it might be possible to start vaporizing the fluid before the EGX if a working fluid with a lower boiling temperature is chosen. This might enhance mass/volume flow and thus increase the efficiency of the expander. Higher mass/volume flows also result in a higher turbulence within the EGX, which might help to increase heat transport and might balance the temperature distribution inside the EGX. This could be advantageous in terms of thermal stress and durability. [5]

Any of the presented additional heat sources has benefits in terms of power output. They also have impacts on the condenser heat flow. The partial and full flow recuperators reduce the condenser heat flow, the preheater and turbocharger increase it. Higher condenser heat flows are normally considered disadvantageous as they increase the load on the coolant circuit, however they can also be beneficial in cold start to heat up components of the car. The additional weight and cost cannot be assessed at this point. All the different advantages and disadvantages should be pondered in terms of cost effectiveness and CO_2-reduction in an economical reflection.

4.3 Influence of Engine Operating Parameters

With the basics of the influences of system operating parameters and additional heat sources covered, the influence of engine operating parameters was investigated. For add-on systems like the one used in this work, there are three physical connections between the engine and the WHR-system: the coolant circuit (which will be discussed in chapter 5.2), the power feedback (which is neglected for these investigations, but will be discussed in chapter 5.1.6) and the exhaust system. The two most important values for the WHR-system in this context are the temperature and the mass flow of the exhaust gas which together form the EG heat flow. These values depend on the engine operating point and the parameter setting of the engine. For SI engines higher speeds and loads always lead to higher temperatures and higher mass flows which would result in higher power output of the WHR-system. Lesser known is the effect of changes in engine operating parameters. Different configurations lead to different EG values with different influences on the engines efficiency and the emissions of the engine. The target of the following investigations was to find configurations that offer higher power output of the WHR-system without diminishing the overall system efficiency or the emission levels of the engine.

The tests were performed on a WHR-system configuration with an electric preheater which is illustrated in figure 4.24. This configuration (the same as in chapter 4.2.2) was chosen to gain information on the maximal possible power output of the system for any change in the engine operating parameters. The electric power input was not taken into account as the preheater could also be supplied by the engine coolant, but by using an electric preheater any influence on the engine could be omitted. The standard operating parameter range is highlighted in gray, as far as a standard can be defined. Parts of these results have already been published [75], [73].

4.3.1 Coolant Temperature

Configuration: Basic+Preheater OP: 50 km/h $T_{Cond} = 65\,°C$ $p_{low} = 0.75\,bar$ $x = 0.8$

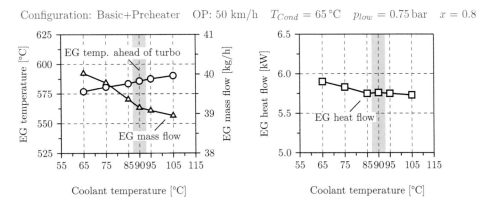

Figure 4.47: Influence of engine coolant temperature on exhaust gas temperature and mass flow

Figure 4.48: Influence of engine coolant temperature on exhaust gas heat flow (in relation to the ambient temperature; calculated after the catalyst)

The coolant temperature has a direct impact on the wall temperature of the combustion chamber and thus on the wall heat flow. Higher temperatures lead to less heat loss, lower oil viscosity, less friction and thus higher engine efficiency, but it also influences material durability, maximal charge and emissions [18]. Because of that, most cars run at a coolant outlet temperature of around 90°C. In order to set the coolant temperature, the thermostat of the engine was deactivated and the coolant temperature was regulated externally. The target temperature was measured at the engine coolant inlet and was controlled with a precision of $\pm$ 1 K. [73]

The exhaust gas temperature ahead of the turbocharger is increased by higher coolant temperatures (figure 4.47). This is most likely due to reduced heat losses in the exhaust port, reduced heat loss through the walls of the combustion chamber and heated-up intake air. The exhaust gas mass flow is lowered with increasing coolant temperature. This is a result of the improved efficiency, which can be attributed to the lowered wall heat transfer and lowered oil viscosity. With a constant air-fuel equivalence ratio ($\lambda = 1$) the lowered fuel consumption results in a lowered charge mass and thus exhaust gas mass flow.

The overall exhaust gas heat flow, which is a product of the mass flow and the difference in temperature-dependent heat capacity, is reduced through higher coolant temperatures (figure 4.48). The increase in temperature cannot compensate the loss in mass flow. [73]

Configuration: Basic+Preheater   OP: 50 km/h $T_{Cond} = 65\,°C$ $p_{low} = 0.75\,\mathrm{bar}$ $x = 0.8$

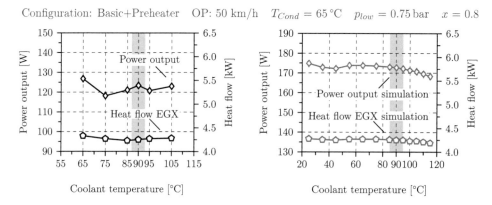

Figure 4.49: Influence of engine coolant temperature on EGX heat flow and power output [73]

Figure 4.50: Simulation results for the influence of engine coolant temperature on EGX heat flow and power output [73]

Figure 4.49 displays the behavior of the power output and the EGX heat flow. The change in the overall heat flow of the exhaust gas causes a reduction of the EGX heat flow and thus a reduction of the power output of the system. The impact on power output is small and almost overshadowed by other measurement uncertainties, yet a trend is visible. [73]

The simulation results also show very small impact of the coolant temperature, but they allow for a more certain conclusion and proof that higher coolant temperatures lead to lower power outputs. This trend could be observed for coolant temperatures from 15-115°C. The simulations reveal alteration efficiencies from $\eta_{Alteration} = 0.1$ % to 0.7 % which is far from the reasonable range. This means the coolant temperature should always be as high as possible (in terms of engine efficiency) and is not to be altered for the WHR-system. [73]

4.3.2 Charge Air Temperature

In turbocharged cars a charge air cooler (CAC) is used to cool down the temperature of the compressed air after being heated up through the compression process. The purpose of this is to increase the density of the air which helps to increase the maximum cylinder charge and to decrease the outlet temperature of the exhaust gas [50]. This means higher charge air temperatures should raise the exhaust gas temperature, which might increase the exhaust gas heat flow and the power output of the WHR-system. For the following tests the charge air temperature was controlled by a water cooled charge air cooler with a precision of ± 1 K. The charge air temperature was measured after the compressor ahead of the inlet manifolds. The 120 km/h-OP was chosen to guarantee a certain level of charge air temperature. [73]

Configuration: Basic+Preheater   OP: 120 km/h $T_{Cond} = 65\,°C$ $p_{low} = 0.75\,bar$ $x = 0.8$

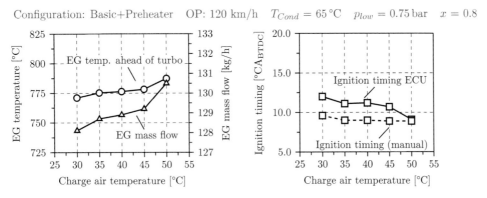

Figure 4.51: Influence of charge air temperature on exhaust gas temperature and mass flow (ignition timing by the ECU)

Figure 4.52: Influence of charge air temperature on ignition timing [73]

Configuration: Basic+Preheater   OP: 120 km/h $T_{Cond} = 65\,°C$ $p_{low} = 0.75\,bar$ $x = 0.8$

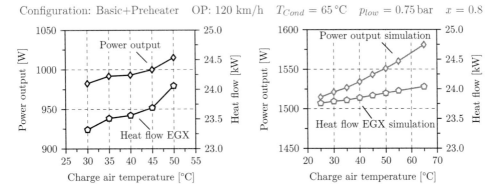

Figure 4.53: Influence of charge air temperature on EGX heat flow and power output (ignition timing by the ECU) [73]

Figure 4.54: Simulation results for the influence of charge air temperature on EGX heat flow and power output

According to figure 4.51 an increase in charge air temperature promotes higher exhaust gas temperatures and mass flows. This is due to two reasons. The first one is the direct influence of charge air temperature and the second one is a change in ignition timing as depicted in figure 4.52. If the ignition timing is controlled by the ECU, the ECU will try to keep a certain distance to the knock limit by adjusting the ignition timing towards later timings. This is done as higher charge temperatures increase the risk of knocking which might damage the engine. Later ignition timing leads to later timing of the center of combustion which reduces peak temperatures and pressures which lowers the risk of knocking, but also lowers the thermal efficiency of the combustion. [73]

For low-load point without the risk of knocking there is another reason for the ECU to alter the ignition timing: the thermodynamic optimal center of the combustion. This is the point in time where 50 % of the fuel is burned. It should be around 7-8°CA after the top dead center (TDC) to reach maximal efficiency. With higher charge temperatures this point wanders towards the TDC due too quicker combustion. In order to keep the center of the combustion constant the ECU has to adapted the ignition timing towards "late".

To exclude the effect of later ignition timing, the variation was repeated with manual timing at the latest ignition time point to prevent knocking at all charge air temperatures. [73]

For points where ignition timing is controlled by the ECU, the gains in exhaust gas temperature and mass flow lead to improved EGX heat flows and thus higher power output. The EGX heat flow is increased by 0.7 kW and power output by 32 W in the course of a charge air temperature rise from 30°C to 50°C. The simulations in figure 4.54 verify this correlation and stretch the assertion to areas from 25°C to 65°C. [73]

Configuration: Basic+Preheater   OP: 120 km/h $T_{Cond} = 65\,°C$ $p_{low} = 0.75\,bar$ $x = 0.8$

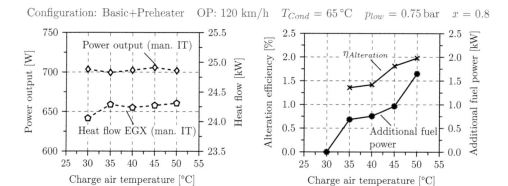

Figure 4.55: Influence of charge air temperature on EGX heat flow and power output (manual ignition timing)

Figure 4.56: Influence of charge air temperature on alteration efficiency (ignition timing by the ECU)

With constant ignition timing, the ignition delay is shortened due to the higher charge air temperatures. According to the simulations this leads to shorter combustion duration and earlier timing of the center of combustion. This in itself could potentially increase the thermal efficiency, but is diminished by higher gas temperatures and thus higher wall heat flows. Overall the engine efficiency is still decreased. [73]

Figure 4.55 shows that the influence of charge air temperature alone is not as large as in combination with the ECU adjusted ignition timing. Overall the EGX heat flow is higher in case of the chosen late ignition timing and does only increase by 0.3 kW with higher charge temperatures. The change in power output is below the standard deviation.[8]

---

[8]The measurements with constant ignition timing were performed with another more used up expander, thus the lower overall power output

Even though the power output of the WHR-system can be enhanced by increasing the charge air temperature, this advantage comes at the cost of additional engine fuel consumption. As figure 4.56 shows this additional fuel consumption translated into fuel power exceeds the additional power output. The alteration efficiency reaches values of 1.4-2 % which does not justify the change in charge air temperature. It should not be altered and should be kept as low as possible for high overall system efficiency and to prevent knocking and emission issues. [73]

4.3.3 Injection Pressure

In direct injection (DI) spark ignition (SI) engines the injection pressure is mostly used to reduce emission levels and at a lesser level to increase the combustion efficiency. Higher injection pressures lead to higher droplet velocity and shorter injection durations which improves fuel homogenization and reduces the generation of particulate emissions [74], [91]. Fuel consumption is only affected by very low injection pressures[9]. To investigate the influence on the WHR-system a variation was performed wherein the injection pressure was altered from 40 bar to 200 bar by setting the desired target value in the ECU. [73]

Configuration: Basic+Preheater   OP: 50 km/h $T_{Cond} = 65\,°C$ $p_{low} = 0.75\,\text{bar}$ $x = 0.8$

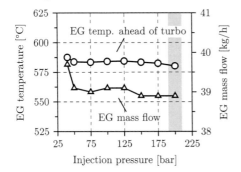
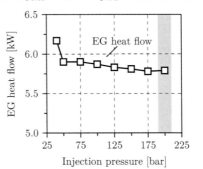

Figure 4.57: Influence of injection pressure on exhaust gas temperature and mass flow [73]

Figure 4.58: Influence of injection pressure on exhaust gas heat flow (in relation to the ambient temperature; calculated after the catalyst)

The engine experiments show that exhaust gas temperature and mass flow are increased by lower injection pressures and as expected the effect is slight and mostly noticeable at pressures below 50 bar. The impact on the exhaust gas heat flow is similar in shape to the temperature and mass flow profile in figure 4.58. [73]

Due to the increased exhaust gas heat flow for lower injection pressures, the EGX heat flow is also increased which leads to higher power outputs (figure 4.59). The power output is increased by 17 W for 40 bar injection pressure instead of the standard pressure of 200 bar. In figure 4.60 simulation results are presented validating the behavior. [73]

In conclusion it can be said that the influence on power output is slight. The measurements show alteration efficiencies between $\eta_{Alteration} = 0{,}1$ % bis 1.9 %, which are below the reasonable threshold. The augmentation comes at the expense of higher fuel consumption and more relevant

---

[9]Depending on the engine operating point and injection system ~around 100 bar

at higher emission levels which most likely cannot be accepted. Thus the injection pressure should be as high as required for the standard engine operation an not lowered when a WHR-system is in use. [73]

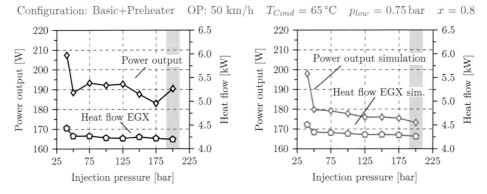

Figure 4.59: Influence of injection pressure on EGX heat flow and power output [73]

Figure 4.60: Simulation results for the influence of injection pressure on EGX heat flow and power output

4.3.4 Air-Fuel Equivalence Ratio

Most modern gasoline engines are operated with a stoichiometric air-fuel equivalence ratio (lambda λ). The possibility of simultaneous oxidation of hydrocarbons (HC), carbon monoxide (CO) and reduction of NO_x within the catalyst is the reason for this. An adjustment of this level in order to gain additional power output from the WHR-system is unlikely due to emission regulation reasons. Yet there are some cases where lambda is altered from its standard value. Two examples are cold start strategy and full load enrichment. One uses a rich mixture to improve the combustion stability and the other uses additional fuel to reach maximal power output while keeping exhaust gas temperatures below certain safety limits. To investigate the behavior of the system in such cases a variation was conducted where lambda was set at 0.85 to 1.15 using 0.05 steps. [73]

This was carried out by the ECU by adjusting the fuel mass flow. The test bench automatically adjusted the torque demand accordingly, to keep the torque torque output constant - thus reducing the air mass flow and the fuel mass flow to the correct ratio. The variation was performed once with automatic ignition timing by the ECU and once with constant manual ignition timing, where the timing was kept at the baseline value for stoichiometric lambda.

Lambda has a distinct impact on exhaust gas temperature and mass flow as can be seen in figure 4.61. The mass flow is increased by leaner mixtures while the temperature has a clear peak at lambda values of $\lambda = 1$. This behavior is steady along the exhaust pipe with the temperature constantly falling due to further heat losses along the way. Only for stoichiometric lambda values it is possible to observe a small gain in exhaust gas temperature after the catalyst. The conversion of HC and CO emissions gifts additional heat. The behavior of the mass flow can be explained through the additional air that is needed to dilute the mixture. This also leads to lower temperatures for lean mixtures. The drop in temperature for richer mixtures can be explained through the additional fuel that has to be vaporized. The vaporization of the fuel cools the exhaust gas. [73]

Configuration: Basic+Preheater   OP: 50 km/h $T_{Cond} = 65\,°C$ $p_{low} = 0.75\,bar$ $x = 0.8$

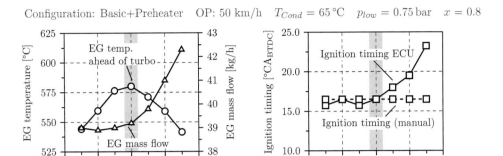

Figure 4.61: Influence of lambda on exhaust gas temperature and mass flow (ignition timing by the ECU) [73]

Figure 4.62: Inluence of lambda on ignition timing

This behavior is identical for constant ignition timing, but the temperature in the lean region is even a bit higher (max. 12 K) due to the later center of combustion. This can be derived from the differences between manual and automatic (by the ECU) ignition timing shown in figure 4.62. The combustion duration is elongated by the lean mixture, which results in a late-shift of the center of combustion. The ECU most likely changes ignition timing to earlier timings to keep the center around the same timing. [73]

Configuration: Basic+Preheater   OP: 50 km/h $T_{Cond} = 65\,°C$ $p_{low} = 0.75\,bar$ $x = 0.8$

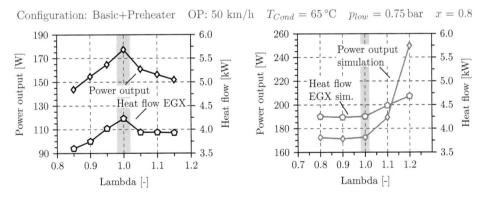

Figure 4.63: Influence of lambda on EGX heat flow and power output (ignition timing by the ECU) [73]

Figure 4.64: Simulation results for the influence of lambda on EGX heat flow and power output

The EGX heat flow resulting from the exhaust gas temperature and the mass flow curves that are shown in figure 4.63 present a clear peak at the lambda value of 1. This peak also translates into a maximum power output at the same point. The increase in mass flow for leaner mixtures is not able to compensate the loss in temperature which results in a clear optimum for lambda. [73]

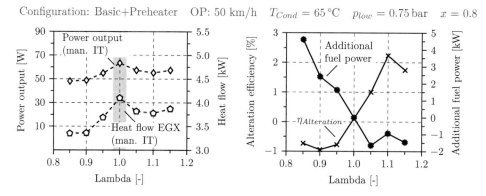

Figure 4.65: Influence of lambda on EGX heat flow and power output (manual ignition timing)

Figure 4.66: Influence of lambda on alteration efficiency (ignition timing by the ECU)

The simulation results depicted in figure 4.64 cannot represent the same behavior. The temperature and mass flow curves of the exhaust gas show a similar trend (not depicted), but the calculated values for mass flow are too high in the lean area and the influence on temperature is calculated too low. In combination with the simulation model that is calibrated for lambda values around 1, the simulation delivers incorrect values. The power output is not reduced enough for rich values and is too high for lean values. To gain more precise results, the simulation model of the engine would have to be fine tuned for different lambda values and the simulation model of the WHR-system would need to have more realistic exhaust gas composition input values and an extended EGX map.

The basic system performance is not changed by the manual ignition timing as can be observed in figure 4.65. There is still a clear optimum of EGX heat flow and power output at a lambda value of $\lambda = 1$. The influence of lambda exceeds the influence of earlier ignition timing in these investigations, most likely due to the ECU only trying to keep the center of the combustion constant by changing the ignition timing. [10] [73]

As lambda values around $\lambda = 1$ show the best results concerning power output, there does not seem to be any potential in changing it for increased system efficiency. Any alteration always comes with a drop in WHR-system power output. The fuel consumption of the engine is lower for leaner mixtures which is the reason for the increased alteration efficiency depicted in figure 4.66. But this comes at the cost of additional NO_x emissions which cannot be reduced by the catalyst during lean operation. The conclusion is that lambda has a clear influence and that any other lambda value than 1 is not beneficial for the WHR-system. [73]

4.3.5 Injection timing

Injection timing is a crucial lever on exhaust gas emissions of gasoline engines with direct injection. It is mainly controlled by the start of injection (SOI). Too early injection timing may lead to piston wetting which results in higher PM/PN and HC emissions [107]. If the timing is too late there is too little time for homogenization which also results in higher PM/PN emissions as

---

[10]The measurements with constant ignition timing were performed with another more used up expander, thus the lower overall power output

well as worse fuel economy [74]. Thus injection timing has an impact on the exhaust gas heat flow. In order to gain information about the influence on the WHR-system, the injection timing was set in 10°CA steps ranging from 235.5 °CA$_{BTDC}$ to 295.5 °CA$_{BTDC}$. [73]

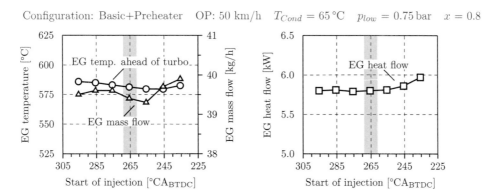

Configuration: Basic+Preheater   OP: 50 km/h $T_{Cond} = 65\,°C$ $p_{low} = 0.75\,bar$ $x = 0.8$

Figure 4.67: Influence of injection timing on exhaust gas temperature and mass flow

Figure 4.68: Influence of injection timing on exhaust gas heat flow (in relation to the ambient temperature; calculated after the catalyst)

Exhaust gas temperature and mass flow show a minimum near the standard SOI (around 265°CA$_{BTDC}$) for this operating point, as indicated in figure 4.67. Before and after this point the temperature rises due to insufficient homogenization caused by time constraints and piston wetting. The mass flow is also affected by these disadvantages. [73]

The short incline between 285°CA to 295°CA cannot be explained without knowledge of the in-cylinder pressure and is not visible in the simulation results. Figure 4.68 displays a small advantage in EG heat flow for later injection timings which is achieved at the cost of an increase in fuel consumption.

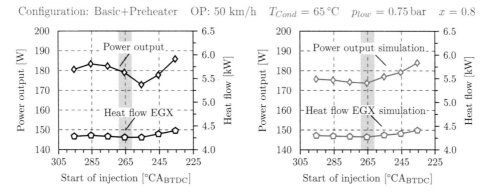

Configuration: Basic+Preheater   OP: 50 km/h $T_{Cond} = 65\,°C$ $p_{low} = 0.75\,bar$ $x = 0.8$

Figure 4.69: Influence of injection timing on EGX heat flow and power output [73]

Figure 4.70: Simulation results for the influence of injection timing on EGX heat flow and power output

The behavior of the EGX heat flow in figure 4.69 complies with the exhaust gas heat flow. Yet the curve of the power output of the WHR-system has a similar shape as the exhaust gas mass flow curve from figure 4.67, which implies that in this case the mass flow has a bigger impact on the power output than the temperature change. There is still a minimal power output close to the standard SOI. The power output is enhanced for earlier and later injection timings with a maximum increase of 6 W in relation to the standard SOI. The simulation results in figure 4.70 verify this behavior even though the incline between 285°CA to 295°CA is not visible. A change in injection timing seems not reasonable taking the small gain in power output and the high impact on exhaust gas emissions into account. [73]

4.3.6 Wastegate Position

The development in gasoline engine moves towards smaller displacement and higher charge pressures. This so called "downsizing" trend demands engines to have some kind of charging system. Most car developers choose a turbocharger in combination with a wastegate for pressure control. An open wastegate (0 %) means all exhaust gas flows through the wastegate and (almost) none through the turbocharger, thus no charge pressure is built up. If the wastegate is closed (100 %) all of the exhaust gas is used by the turbocharger to build up charge pressure. [73]

As passing through the turbocharger produces higher exhaust gas back-pressure, which decreases charge exchange efficiency, it is only used when the resulting charge pressure is needed. This is for example the case for high engine loads. But even for low loads that do not need any charge pressure, it is possible that the wastegate is still kept partially shut to produce a certain charge pressure reserve, which can be used in case of sudden torque demand. This charge pressure reserve comes with a loss of exhaust gas enthalpy, as the exhaust gas is expanded and thereby cooled which lowers its heat flow. This heat flow could potentially be used by the WHR-system to increase its power output. To examine this potential tests with various wastegate positions were performed. The wastegate position was manipulated by setting the respective value in the ECU. The torque was kept constant by restraining the charge pressure with the throttle. [73]

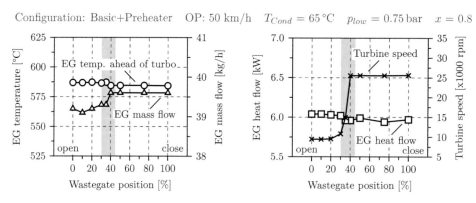

Configuration: Basic+Preheater   OP: 50 km/h $T_{Cond} = 65\,°C$ $p_{low} = 0.75$ bar $x = 0.8$

Figure 4.71: Influence of wastegate position on exhaust gas temperature and mass flow

Figure 4.72: Influence of wastegate position on turbine speed and exhaust gas heat flow (in relation to the ambient temperature; calculated after the catalyst) [73]

By closing the wastegate the exhaust gas temperature is lowered, but at the same time the mass flow is increased, as demonstrated in figure 4.71. There is a distinct unsteadiness at the standard wastegate position around 38 %. As shown in figure 4.72, this is the position where the turbine speed is no longer increased by further closing the wastegate. It is also visible, that the exhaust gas heat flow is lowered for a more closed wastegate, which was to be expected. [73]

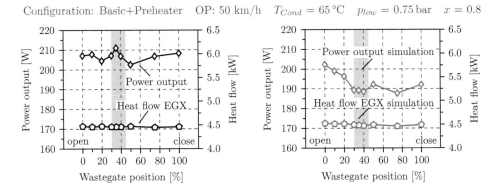

Configuration: Basic+Preheater   OP: 50 km/h $T_{Cond} = 65\,°C$ $p_{low} = 0.75\,\text{bar}$ $x = 0.8$

Figure 4.73: Influence of wastegate position on EGX heat flow and power output [73]

Figure 4.74: Simulation results for the influence of wastegate position on EGX heat flow and power output

The impact on the EG heat flow is not as large as expected. This can be explained by taking the ECU's automatic reaction into account. When the engine has to provide additional exhaust gas back-pressure and thus enthalpy flow, its efficiency is reduced and thus it has to consume more fuel and air to keep the torque output constant. This results in additional mass flow of exhaust gas. If the engine does not have to provide the back pressure, the potential enthalpy of the charge pressure reserve is not available for disposal; the engine just runs at a higher efficiency and thus reduces most of its additional exhaust gas back-pressure and thus enthalpy flow. [73]

Due to lower mass flows and higher temperatures at more open wastegate positions, most of the gain in exhaust gas heat flow is lost on its way downstream of the exhaust pipe to the EGX. That is why the EGX heat flow is mostly unaffected by the changes in the wastegate position as is visible in figure 4.73. The power output is not altered significantly, which is to be expected due to the constant EGX heat flow. [73]

The simulation results (figure 4.74) show small increases in EGX heat flow and power output for more open wastegate positions. This is due to the underrepresented heat losses in the exhaust pipe based on the combination of smaller mass flows and higher temperatures.

The conclusion is that the recovery of the charge pressure reserve provides an increase in exhaust gas heat flow, but due to losses within the exhaust pipe it is practically lost again until reaching the exhaust gas heat exchanger. Thus there is no noteworthy change in the WHR-system's power output. The overall system efficiency is still increased through opening the wastegate which improves charge exchange efficiency, but this is in conflict with the torque characteristics of the engine which is why the charge pressure reserve is kept in the first place. [73]

4.3.7 Ignition Timing

As mentioned before, the ignition timing (IT) does have a major impact on engine efficiency and exhaust gas heat flow. The reason for this is the influence of ignition timing on the location of the center of combustion. If the ignition is set too early, too much heat is lost via the cylinder walls before the gas can expand; and if it is set too late, the peak temperatures are lowered and the time for expanding the combustion gas is too short which both results in thermodynamic losses. These losses lead to higher exhaust gas temperatures and mass flows which translates into higher heat flows. Shudo [93] proofed this in his work and also hinted that using a WHR-system some of these losses can be recovered. This could be interesting for operating points in which the ignition timing is set late on purpose, e.g. catalyst heating strategies or full load operation. [73]

To investigate this potential a variation of ignition timing was performed, with the ignition being incrementally set to "late" by setting the respective ECU value. As the engine did not have any kind of indication system for in-cylinder pressure the variations were mostly performed towards later ignition timing in relation to the standard timing. This approach was chosen to prevent the engine from knocking. Earlier ignition timings were investigated using the simulation model (3.3). [73]

Figure 4.75 shows that by setting the ignition towards later timings, both exhaust gas temperature and mass flow are vastly increased. For the single ignition timing earlier than the standard timing, they both decrease, as the engine efficiency is probably enhanced. Presumably this timing is not used as the standard timing to keep a higher security distance towards the knock limit. [73]

Configuration: Basic+Preheater   OP: 50 km/h $T_{Cond} = 65\,°C$ $p_{low} = 0.75$ bar $x = 0.8$

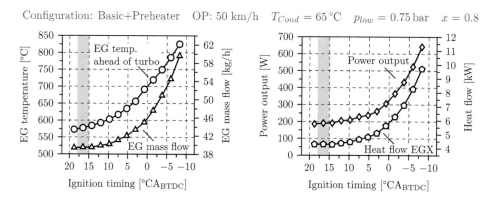

Figure 4.75: Influence of ignition timing on exhaust gas temperature and mass flow

Figure 4.76: Influence of ignition timing on EGX heat flow and power output [73]

Higher exhaust gas temperatures and mass flows lead to higher heat flows and thus higher EGX heat flows, as can be seen in figure 4.76. The EGX heat flow is increased by 5,4 kW which is more than double the base heat flow of 4.3 kW. As the efficiency of the WHR-system increases (3 % to 5 %) with heat input, the power output is increased from 190 W to 642 W which is around 3.4 times more power. This illustrates that ignition timing is a significant lever for power output. It is possible to enhance the output of the WHR-system multiple times by delaying the ignition timing while keeping the engine torque constant. The exhaust gas emissions are generally lowered by lower peak combustion temperatures and higher exhaust gas temperatures. Only PN/PM emissions can happen to be increased with later ignition timing. [73]

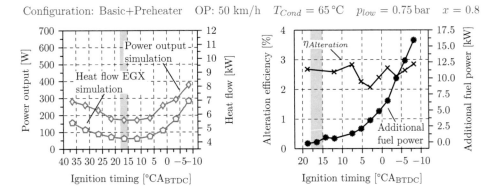

Figure 4.77: Simulation results for the influence of ignition timing on EGX heat flow and power output [73]

Figure 4.78: Influence of ignition timing on alteration efficiency and additional fuel consumption [73]

The simulation results shown in figure 4.77 verify the previous statements and also show that the same statements can be made for ignition timings earlier than the thermodynamic optimal timing. As efficiency is decreased, the EGX heat flow is increased through higher exhaust gas temperatures and mass flows which in the end results in higher WHR-system power output. The disadvantages of earlier timings are the risk of knocking and the tendentially worse exhaust gas emission levels. [73]

Yet from an overall system efficiency standpoint an alteration of ignition timing is not reasonable. The efficiency and power output of the WHR-system are inclined, but the decrease in engine efficiency is too high to keep the overall system efficiency at a constant level. In figure 4.78 a representation of the additional fuel consumption in relation to the base operating point can be seen. The fuel consumption is increased by 54 % from the standard ignition timing to a timing which is 25°CA later. Even though the WHR-system can recover 2.8 % of that energy, it is not enough to compensate the overall system efficiency loss. [73]

But when an alteration of ignition timing is needed (catalyst heating, knock prevention) some of the lost energy can be recovered and the overall system efficiency would be better than without a WHR-system. Another way to look at the results is to take the ignition timing as a possibility to enhance the power output of the WHR-system without changing the vehicle speed. The timing can be set late without the driver noticing any change in torque while the power output can be set to fit any demands. For example this could allow abandoning the generator of the car and relying completely on the electric power provided by the expander generator of which the electric power output can be adjusted to the car's needs by altering the ignition timing. [73]

Ignition Timing and Overall System Efficiency

To give an outlook of the possibilities for improving the overall system efficiency through changing engine operating parameters and thus improving the power output of the WHR-system an example for an alteration of ignition timing will be presented in the following. Ignition timing is not really the driving force for the change in system efficiency but it represents the power that is generated with the WHR-system and the efficiency loss of the engine itself. The electric power output is taken from the measurements depicted in figure 4.76.

Figure 4.79 shows a comparison of system efficiencies for two cases. The first ("WHR-System") produces electric energy by altering ignition timing of the engine while using the resulting heat flow with the WHR-system. And the other ("Generator") is a combination of engine and generator that produces the same amount of electric power as the WHR-system for a given, constant ignition timing. The torque demand of the generator is taken from a generator map and added to the base torque demand (in accordance with 2.2.4). The resulting system efficiency (based on measured fuel consumption: equation 2.17) of both cases is plotted over the ignition timing.

In figure 4.80 another comparison of system efficiency is presented. There are three different cases presented. The first case ("Engine") shows the system efficiency when the ignition timing of an engine without WHR-system is altered. The second case ("WHR-system + electr. Motor") shows the impact of a WHR-system on this setup with the power output being used by an additional electric motor ($\eta = 95\,\%$) that supports the combustion engine (in accordance with chapter 2.2.4). Any kind of feedback loop after the first feedback was neglected due to the high amount of measurements needed, but the influence for small changes in ignition timing should be negligible as will be shown in chapter 5.1.6. The last case shows the same results as the first case without the ignition timing being changed at all.

The WHR-system in figure 4.79 has an advantage over the generator version for small deviations from the standard ignition timing. The reason is that under these conditions the WHR-system produces electric energy practically without additional fuel consumption (cf. 4.78) while the generator needs additional torque to supply the same energy. For later ignition timings this trend is turned around as the alteration efficiency of the WHR-system is between 2 % to 3 % while the generator works at an alteration efficiency of 10,3 % to 14,8 %. Thus the additional fuel is used more efficiently by the generator. For the IT change to be reasonable the alteration efficiency of the WHR-system would have to be as high as that of the generator.

Configuration: Basic+Preheater   OP: 50 km/h $T_{Cond} = 65\,°C$ $p_{low} = 0.75\,bar$ $x = 0.8$

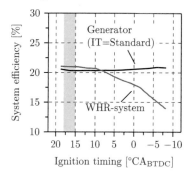

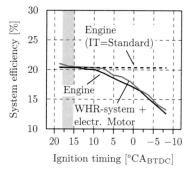

Figure 4.79: Influence of ignition timing on system efficiency for different electric power generation strategies

Figure 4.80: Influence of ignition timing on system efficiency for different propulsion strategies

The use of the WHR-system's power output by an electric motor in figure 4.80 has a small advantage around the standard ignition timing which can be attributed to the free power generation. If the ignition timing is set to later timings the WHR-system with electric motor is still better than just the engine, however the system efficiency declines especially when compared to the engine without changed IT.

The conclusion is, that the alteration of ignition timing is not beneficial for the overall system efficiency. But any system is more efficient with a WHR-system for cases in which the ignition timing has to be changed than without one.

4.3.8 Conclusion

The investigation of the influence of engine operating parameters showed that it is possible to enhance the power output of a WHR-system through suitable parameters. But they also show that it is not reasonable to change any of them to improve the overall system efficiency. The optimal setup is a setup with the highest possible engine efficiency. For the common system development level, the efficiency with which the combustion engine uses the fuel is always higher than one with which the WHR-system would use it. For this condition to change, the alteration efficiency would have to be as high as the efficiency of the engine. This means the efficiency of the WHR-system would have to be much higher than the one of today's systems. Due to constraints in cost, packaging, operating range and mostly temperature levels in comparison to the combustion temperature levels, this is very unlikely to be realizable. Yet it is possible to recover energy losses at points where the parameters have to be changed for specific reasons. Coolant and charge air temperature as well as injection pressure, injection timing and wastegate position did have small impacts on power output. The influence of air-fuel equivalence ratio was considerably larger, but the highest impact was found for the ignition timing. It allowed increasing the output by more than three times at the investigated operating point. This might be useful to give a degree of freedom to influence the power output of the WHR-system without changing the engine speed and torque output. [73]

5 Results for Cold Start Operation

Thanks to the previously obtained insights into the influences of system operating parameters, system configurations and engine operating parameters it is possible to investigate the system behavior during cold start. Boretti predicted improvements in fuel economy based on simulations [8] and in combination with an expander bypass. However, the behavior of real WHR-systems during cold start is widely unknown and almost no results have been published. The basic behavior and possible influences were examined on the test bench to gain real results. After that a final investigation was performed with the optimal system parameters and the aim of finding possibilities to use the remaining condenser heat in a way to improve the engine's cold start and thus improving the overall system efficiency.

5.1 System Behavior during Cold Start

The cold starts were performed according to the VDA220 standard warm up cycle for passenger cars [108]. This driving cycle was created by the VDA to have a cycle with low engine load and low dynamics which helps in representing differences in the warm up of engines and passenger cabins at difficult environmental conditions. The cycle starts with an acceleration to a speed of 50 km/h which is kept for 30 min (end torque: 32.9 Nm, end speed 1558 rpm). After that the engine switches to idle for 15 min. The tests were run at starting temperatures of 0°C and +20°C (results are mostly shown for 0°C). Engine speed and torque were taken from a middle class passenger car (D-segment). The WHR-system was configured with the basic setup consisting of EGX, expander, condenser and pump, if not mentioned otherwise. Parts of these results were also presented in [75].

5.1.1 General Behavior

The first tests were performed to understand the basic system behavior in cold start. The focus was set on the interaction of temperature levels, coolant flow rates and heat flows. The tests were run with a separated cooling circuit (configuration V1) as presented in illustration A.9. The basic investigations were performed at a steam quality of 0.8, a low-pressure of 1 bar, a condenser coolant setpoint[1] of 65°C and an ambient temperature of 0°C (= starting temperature). The system operating parameters were chosen based on the results from chapter 4.1 and their demand on system complexity (additional components etc.).

Figure 5.1 shows the temperature profiles of the working fluid at different points of the WHR-system. The evaluation of the working fluid temperatures at specific points in the system displays the expected differences in steady-state temperature levels. The steam reaches around 140°C after passing the exhaust gas heat exchanger and drops down to around 100°C behind the expander. The condenser cools the remaining steam down to approximately 65°C.

---

[1]The condenser coolant circuit was not externally heated. The only heat source was the condenser. The circuit was isolated and minimized as much as possible.

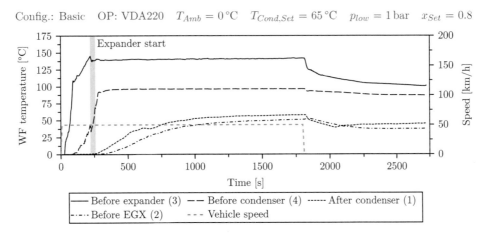

Config.: Basic   OP: VDA220 $T_{Amb} = 0\,°C$ $T_{Cond,Set} = 65\,°C$ $p_{low} = 1\,bar$ $x_{Set} = 0.8$

Figure 5.1: Temperature profile of the working fluid at specific points of the system

Yet there is a vast difference in time until each point reaches its steady-state. The earliest point to reach the steady-state is the point ahead of the expander which takes 210 s. It takes up to 90 s until the temperature downstream the expander starts rising and another 200 s until it achieves steady-state. Shortly before the temperature after the expander arrives at its stationary condition, the expander starts turning and thus providing energy, most likely because it is completely filled with steam at that time. The temperature downstream of the condenser takes about 15 min until it reaches its set point.

These results for the cold start show that the heat entering the system through the exhaust gas heat exchanger reaches one component after the after but only after the former component is heated significantly. This can be visualized by a wave that reaches each component and only travels further after each barrier (temperature level) is surpassed. The first one is the exchanger itself, next is the expander followed by the condenser. After the pump is heated, the temperature ahead of the EGX starts to finally rise. The reason behind this is the relatively small mass flow of hot steam compared to the mass of the components. Depending on the working fluid and the steam quality, the mass flow reaches around 5-15 kg/h while the components weigh between 5-10 kg each. Also their surface area is large (exchangers, scrolls...) meaning they take up a lot of heat in a short time. This is why the working fluid might enter a component hot, but is cooled down to almost starting conditions when it leaves the component.

The start-up behavior of the expander can also be observed by taking a look at the profiles of the steam qualities ahead and after the expander, which are depicted in figure 5.2. The setpoint for the steam quality ahead of the expander (3) is 0.8 from the beginning. This value is reached after 150 s. The first time that steam can be detected after the expander (4) is around 240 s. Shortly before steam after the expander can be detected, it starts to turn. This leads to the conclusion that the steam entering the expander is condensed by its cold walls, as long as the expander is not hot enough. When the expander reaches its target temperature it is filled with steam and starts turning. If the expander really does start at that point in time depends on its current constitution. The amount of oil, the position of the scroll, the state of the ball bearings all seem[2] to influence the start of the expander. To avoid any major deviations through this stochastic correlation a start-up procedure was chosen: if certain values (mainly T_4) were reached the mass flow was suddenly tripled for a second to "kick-start" the expander. This sudden change leads to

---

[2]The influences could not be tested within this project, but the difficulties in start-up due to high rates of leakage are also mentioned by Körner [57].

a variation in back-pressure causing a temporary drop in steam quality ahead of the expander: this is highlighted in gray in figure 5.2. This procedure proofed to be very reliable and showed high reproducibility. After that the controller keeps the steam quality at a steady level until the idle part when it cannot be kept due to the small heat input.

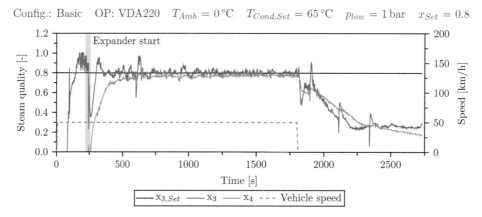

Config.: Basic   OP: VDA220 $T_{Amb} = 0\,°C$ $T_{Cond,Set} = 65\,°C$ $p_{low} = 1\,bar$ $x_{Set} = 0.8$

Figure 5.2: Steam quality profile at specific points of the system

After heating up, the turning expander propels the generator which produces electric energy. The profile of the power output is presented in figure 5.3. The synchronicity of steam after the expander and the start of power generation can be observed by comparing the gray area and the curve of the power output. At the beginning the power output climbs fast but then it takes a while until reaching steady-state. This can be explained through the heat-up of the remaining system components. The power output is quickly reduced in the idle phase. The behavior of the power output is very similar to the profile of the working fluid temperature ahead of the EGX in figure 5.1. This leads to the assumption that in order to improve the system performance in cold start situations two aspects of the system have to be improved. The first is the heat-up of the expander which could be optimized by reducing its mass or by an additional heat source (e.g. coolant); and the second is the temperature level ahead of the EGX, which could be improved by reducing the thermal masses within the system, insulating the components or by using additional heat sources like the ones discussed in chapter 4.2. Kraljevic [58] proposes another approach wherein he uses a steam accumulator to store hot pressurized water which vaporizes when released into the system. This might help improve the start-up of the system, but involves additional components and weight.

Config.: Basic   OP: VDA220 $T_{Amb} = 0\,°C$ $T_{Cond,Set} = 65\,°C$ $p_{low} = 1\,bar$ $x_{Set} = 0.8$

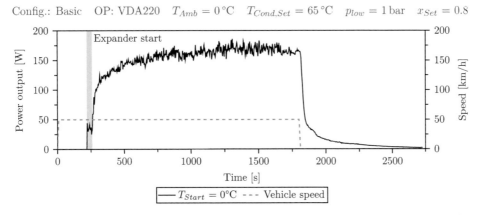

Figure 5.3: Power output profile of the system for a VDA220 at a starting temperature of 0°C. (The curve is jittery due to the mass flow control, variations in expander speed and irregularities during the vaporization)

5.1.2 Influence of Start Temperature

With the basic cold start behavior investigated the next point of interest was the influence of the starting temperature. This was tested with the same system and the same parameter setting as before. The whole WHR-system and the engine were cooled down to four different temperatures (-10°C, 0°C, +10°C, +20°C). The results for the power output are presented in figure 5.4.

Config.: Basic   OP: VDA220 $T_{Cond,Set} = 65\,°C$ $p_{low} = 1\,bar$ $x_{Set} = 0.8$

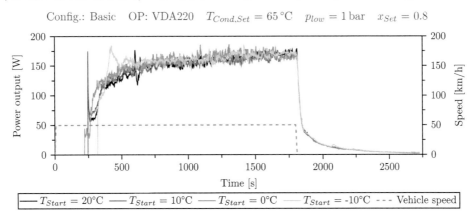

Figure 5.4: Power output profile of the system at different starting temperatures

Except for the remaining stochastic events at the start-up of the expander all four curves are similar. None of them behaves different during the system's heat up, also none shows a difference in steady-state power output. There is also no clear temperature coherence to be observed.

To explain these unexpected results the profiles of the exhaust gas heat flows for the different starting temperatures are displayed in figure 5.5. These were calculated for the exhaust gas values after the catalyst in relation to an ambient temperature of 20°C. If they were calculated in relation to their real ambient temperature, the offset between the curves would be even larger as the temperature difference would increase.

The exhaust gas heat flows show a clear temperature dependency, especially in the beginning. The colder the ambient/starting temperature is, the higher is the exhaust gas flow. The peak within the first 500 s seems to compensate any thermal losses and any additional thermal energy demand through the cold start at lower temperatures. Even after that peak, the heat supply from the exhaust gas is higher for lower starting temperatures. This seems to be the cause for the temperature independence of the power output.

One of the reasons for the elevated exhaust gas heat flows comes from the cold start of the engine. Through lower starting temperatures the engine has to overcome higher wall heat losses and higher friction losses due to higher oil viscosity. This leads to higher fuel consumption resulting in higher exhaust gas mass flows. The temperature level (after the peak in the beginning) is almost unaltered by lower starting temperatures with a small tendency towards higher temperatures for lower ambient temperatures.

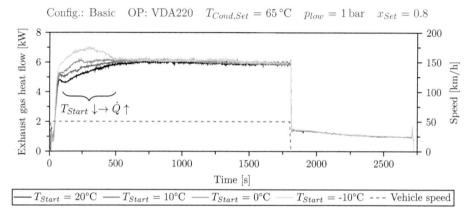

Figure 5.5: Exhaust gas exergy flow at different starting temperatures referred to a ambient temperature of 20°C, calculated after the catalyst

The distinct peak of exhaust gas heat flow at the beginning can be explained based on the connections found in chapter 4.3.7. The ignition timing is set to later timings by the ECU, as can be seen in figure 5.6. Later ignition timing results in an increase of exhaust gas mass flow and exhaust gas temperature. The ECU does this in order to improve the heat-up of the catalyst to secure the reduction of emissions. A byproduct of this "catalyst heating strategy" is the increase of heat flow at colder ambient temperatures that can be used by the WHR-system. Thus the WHR-system is practically independent from starting temperature and is able to recover parts of the engines additional losses during cold start.

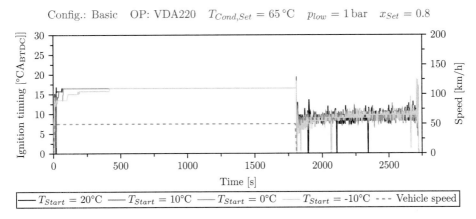

Figure 5.6: Ignition timing for a VDA220 at different starting temperatures (set by ECU)

5.1.3 Influence of Steam Quality

The start-up of the expander is one of the key issues of the cold start of a WHR-system. The target is to reach the operating temperature as quickly as possible. Except for design measures there is also potential in an optimized mass flow and therefore an adapted setpoint for the steam quality. Lower mass flows lead to higher temperatures but higher mass flows improve the efficiency of the EGX (cf. chapter 4.1.1). Thus starting the system with a higher mass flow might help in heating up the components, but in order to reach the desired operating temperature the correct timing is necessary to reduce the mass flow. In order to gain some basic results on this topic four measurements were performed with four different setpoints for steam quality ($x = 0.8, 0.4, 0.2$) and one measurement with a constant mass flow of 60 kg/h ($x = 0$). The setpoints were kept constant over the whole cycle. The resulting mass flows are presented in diagram 5.7.

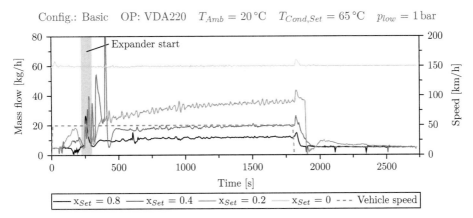

Figure 5.7: Mass flow of working fluid for different set points of steam quality. (Peaks at the expander start are caused by the stochastic variations in the start-up behavior of the expander)

Setting the desired steam quality to lower values allowed increasing the mass flows in the beginning. Yet for the first 70 s the three measurements with 0.8, 0.4 and 0.2 have the same mass flow, which is the minimal value of 5 kg/h. Only after each one reaches its setpoint the controller increases their mass flow. After the expander start the mass flows are controlled according to the steam quality setpoints.

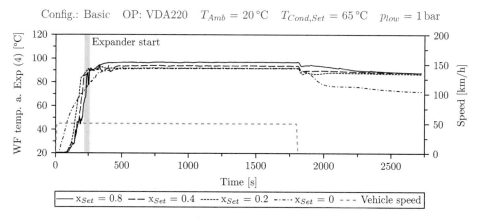

Figure 5.8: Working fluid temperatures after expander (4) for different set points of steam quality

As a characteristic for the heat-up of the expander, the outlet temperature of the working fluid after the expander was chosen. The influence of increased mass flow / reduced steam quality on it is depicted in figure 5.8. Higher mass flows do indeed improve the heat-up before expander start. Yet at the moment that the expander is kickstarted the temperatures are almost identical. The measurement with a mass flow of 60 kg/h even has a lower temperature at that time even though it showed clearly higher values after the start of the driving cycle. Another observation that was made, is that higher mass flows lead to an earlier arrival of the "heat wave" at the condenser and the remaining components. The wave is quicker but much more flattened. This leads to earlier heat losses.

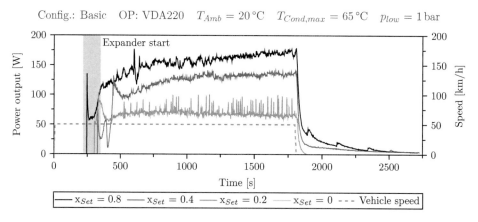

Figure 5.9: Power output for different set points of steam quality

Figure 5.9 presents the outcome for the power output. The power output after the expander start is reduced through the higher mass flows / lower steam qualities, which was to be expected when considering the steady-state results (c.f. chapter 4.1.1). Despite the earlier heat-up of the expander, no systematic advantage in start-up behavior can be found. The mass flow must be reduced at the right moment to lift temperature and steam quality to the right levels before the expander is started. Otherwise the start-up behavior is practically unfazed and still mainly underlies stochastic deviations. This was not in the focus of the work and there no such strategy could be developed and the standard procedure with a fixed setpoint was kept for the remaining investigations.

5.1.4 Influence of Low-Temperature Level

The influence of the low-temperature level on the system and its power output in steady state was already investigated in chapter 4.1.2. In this chapter its impact on the cold start of the system will be determined. The low-temperature is mainly set by the condenser coolant temperature level, which is controlled by the heat rejection of the condenser radiator. No external heat was added to the condenser, only the setpoint for the temperature controller was altered. The results for the working fluid temperature before the EGX are presented in figure 5.10.

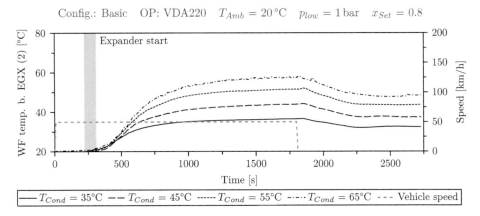

Figure 5.10: Working fluid temperatures before EGX for different condenser coolant temperature levels

The temperature profiles are practically identical for the first 500 s. At this point in time the "heat-wave" reaches the condenser and the controller starts cooling down the condenser fluid. After that the temperatures behave according to the respective setpoint of the controller and reach steady-state one after the other in order of temperature setpoints.

Figure 5.11 displays the power output for each measurement. The start-up behavior of the expander is unaltered by the low-temperature setpoint. The identical temperature profiles within the first 500 s are the reason for this. After that point in time the curves diverge and each temperature setpoint delivers its specific power output. Again these develop according to the temperature profiles; higher condenser coolant temperatures achieve higher power outputs.

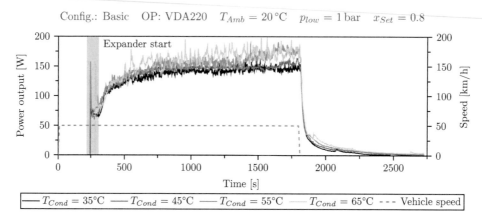

Figure 5.11: Power output for different condenser coolant temperature levels

5.1.5 Influence of Low-Pressure

The last system operating parameter that was investigated within this project was the low-pressure. Low-pressure has a significant impact on the performance of the expander in steady-state. Inappropriate low-pressures, especially if they are too high, lead to unideal pressure ratios over the expander which reduces its power output (cf. chapter 4.1.3). In contrast, higher low-pressures also result in higher temperature levels after the expander (4) which might help heating up the WHR-system. The potential influence on the cold start behavior was investigated for three different low-pressures (0.75 bar, 1 bar, 1.25 bar). The resulting pressure ratios are depicted in diagram 5.12.

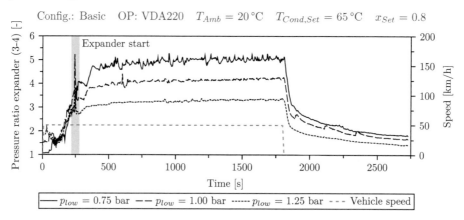

Figure 5.12: Pressure ratio over the expander (3-4) for different low-pressure levels

The measurements confirm that smaller low-pressures produce higher pressure ratios even in non steady-state operation. This is distinctly apparent after the expander started. In the first 120 s the pressure ratios seem to be inflicted by stoichiometric deviations of the expander position and lubrication. After that the first steam is produced and the ratios equalize. After the expander start they slightly incline until reaching steady-state shortly before the idle phase sets in.

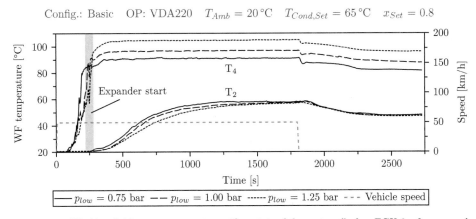

Figure 5.13: Working fluid temperatures at specific points of the system (before EGX 2; after expander 4) for different low-pressure levels

Figure 5.13 displays the profiles of the temperatures before the EGX (2) and after the expander (4) for the three different low-pressures. At first the profiles of T_4 behave similarly except for the time around the start-up of the expander, which can be attributed to the stochastic variations. After that the higher low-pressures result in higher working fluid temperatures after the expander. Yet the temperatures T_2 scale inverse to that. Higher low-pressures lead to lower temperatures before the EGX. This can be explained by the additional heat rejection over the condenser and the increased heat losses over the walls of the system that are caused by the higher values of T_4. Before the end of the 50 km/h phase all three curves reach steady-state at the same level. In conclusion smaller low-pressures are favorable considering the pressure ratio over the expander and also for a quicker system heat-up.

If a recuperator is installed within the WHR-system the previous statement must be amended. The recuperator would transfer more heat from the point after the expander (4) to the point before the EGX (2) for higher levels of T_4, which is the case for higher low-pressures.

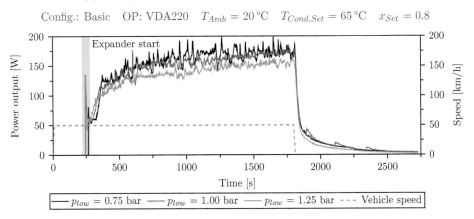

Figure 5.14: Power output of the system for different low-pressure levels

The power output resulting from the three different low-pressures is depicted in figure 5.14. The power output curves fit the behavior of the pressure ratios and the working fluid temperature T_2. Low-pressures that cause pressure ratios close to the ideal ratio produce higher power outputs.

The start-up of the expander is not visibly changed and the differences in power output shortly after it are marginal.

A visible side effect is the more pronounced fluctuation of the power output for lower low-pressures. This can be attributed to the smaller distance to the boiling temperature that is resulting from the constant condenser coolant temperature. The mass flow tends to fluctuate more as a result of the reduced subcooling and the control system has to accommodate for that. The result is an imperfect mass flow and a divergence in steam quality.

5.1.6 Influence of Power Feedback

So far the generated power from the WHR-system is not fed back to the engine / drivetrain in any way. A reduction in demanded driving load would result in a decrease of exhaust gas heat flow which would reduce the power output (and the cycle would continue). An experiment was conceived to take a look at the possible decline in WHR-system performance through the power feedback. In order to do so without any real physical feedback, the method from chapter 2.2.4 was chosen. The origin of the fed back power was the power output curve from an earlier measurement without power feedback at the same conditions. This curve was smoothed with a moving average over 30 s to avoid unsteadiness. The combined efficiency of AC-converter and electric motor was estimated to be 90 %. This gave a new curve for the demanded torque which is presented in figure 5.15.

Config.: Basic   OP: VDA220 $T_{Amb} = 20\,°C$ $T_{Cond,Set} = 65\,°C$ $p_{low} = 0.75\,bar$ $x_{Set} = 0.8$

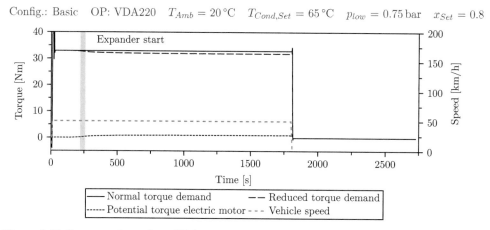

Figure 5.15: Set torque demand in a VDA220 with and without the support of an electric motor driven by the produced power output from the WHR-system

The potential torque that could be produced by an electric motor is zero at the start of the cycle and starts increasing after the expander has started. In the end it reaches just above 1 Nm which reduces the normal torque demand at the end of the cycle from 32.9 Nm down to 31.9 Nm (around 3.04 %).

The power outputs from both measurements with and without power feedback are shown in figure 5.16. Through the feedback the power output is reduced, but the reduction is slight. After 1800 s 67.8 Wh are produced instead of 71.0 Wh (-4,5 %). The potential torque that is subtracted would only be altered insignificantly making the results reliable.

Config.: Basic   OP: VDA220 $T_{Amb} = 20\,^\circ\text{C}$ $T_{Cond,Set} = 65\,^\circ\text{C}$ $p_{low} = 0.75\,\text{bar}$ $x_{Set} = 0.8$

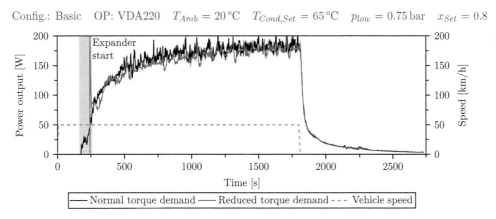

Figure 5.16: Comparison of the power output within a VDA220 with and without the support of an electric motor driven by the produced power output from the WHR-system

Config.: Basic   OP: VDA220 $T_{Amb} = 20\,^\circ\text{C}$ $T_{Cond,Set} = 65\,^\circ\text{C}$ $p_{low} = 0.75\,\text{bar}$ $x_{Set} = 0.8$

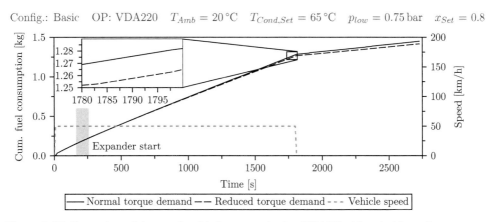

Figure 5.17: Comparison of the cumulated fuel consumption in a VDA220 with and without the support of an electric motor driven by the produced power output from the WHR-system

The reduction in torque demand through the power feedback leads to a reduction of fuel consumption, which is presented as the cumulated fuel consumption in figure 5.17. The consumption is identical up to the point when the expander starts delivering electric power. After that the reduced torque lowers the fuel consumption which results in 1.4 % fuel saving after 1800 s.

5.1.7 Influence of Additional Heat Sources

Condenser Recuperator and Preheater
The previous subchapters covered the influence of system parameters and the general behavior of a WHR-system in cold start. The impact of different system designs will be discussed in the following. These include configurations with a (partial flow) recuperator, a preheater and a turbocharger. The measurements were performed with identical system operating parameters. The preheater was supplied with engine coolant from the cylinder head (not externally as in

chapter 4.2.2). A schematic can be found in appendix A.3. The main focus was the effect on system heat-up and power output. The results for the working fluid temperature before the EGX (2) are presented in figure 5.18.

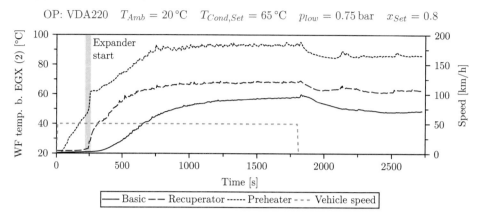

Figure 5.18: Temperature of the working fluid ahead of exhaust gas heat exchanger (2) for different circuit configurations

Compared to the basic configuration without any additional heat sources it is possible to improve the temperature of the working fluid before the EGX considerably by utilizing a recuperator or a preheater. Both enhance the steady-state temperature level and also improve the warm-up phase in terms of start time and duration. Even in the idle phase the temperature levels are elevated. The preheater reaches a temperature level of around 93°C, the recuperator (partial flow) of around 68°C and the basic configuration reaches 58°C. The improvement from the recuperator comes for free as the heat is exchanged within the system, but the preheater heat flow is taken from the engine coolant. This elongates the engine warm-up by additional 133 s. This results in an additional fuel consumption of 6 g which correlates to an increase of 1 %. The impact on exhaust gas temperature and mass flow is not significant.

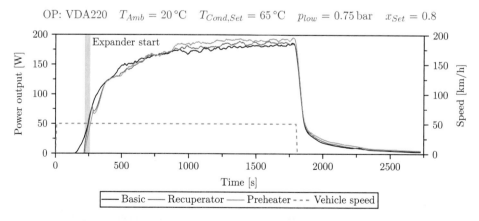

Figure 5.19: Power output for different circuit configurations (floating average over 30 s)

Even though the temperatures inclined earlier and also reach higher levels in the beginning, the power output is only improved when the system reaches steady-state. Figure 5.19 shows that the power output can be improved through the elevated temperatures resulting from the additional heat sources. Yet a distinct gain can only be identified after around 900 s. The moment at which the expander starts up is not altered as well as the power output during the heat up of the system. As the heat-up behavior is not improved by the preheater, it would be advisable to constrain the coolant flow to it in the beginning, which would benefit the engines cold-start and still give an advantage to the WHR-system in steady-state operation. A reason for the unaffected power output in the beginning is presented in diagram 5.20.

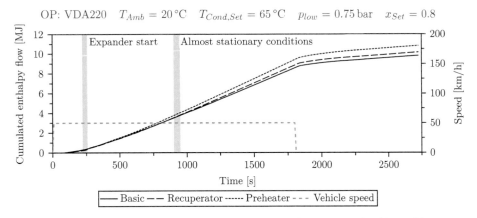

Figure 5.20: Cumulated enthalpy flow for different circuit configurations with additional heat sources (in relation to the ambient temperature)

The power output of the system is mainly influenced by the operating temperature of the expander. To heat the expander up from 20°C to 140°C an energy amount of around 1 MJ is needed. Figure 5.20 shows the cumulated enthalpy flow that is delivered to the expander by the working fluid. The needed enthalpy amount is delivered after approximately 500 s, but as the expander heats up the needed temperature difference inclines and less of the delivered heat can be used to increase its temperature. Thus it takes much more time to reach steady-state in the end.

The additional heat sources deliver working fluid at higher temperatures to the EGX. This reduces the EGX heat flow but increases the overall mass flow with a constant steam quality and thus the enthalpy flow (as described in chapter 4.2). In a heat up scenario the additional heat ahead of the EGX does also increase the enthalpy flow, but the difference seems to be marginal for the first 900 s. Thus the heat-up of the expander is not measurably enhanced. Steady-state operation is increased as the maximal enthalpy flow is higher with the recuperator and the preheater.

Turbocharger

The turbocharger is an external heat source that can be used in steady-state without significant disadvantages, as described in chapter 4.2.4. For cold start scenarios it was found that the heat rejection from the turbocharger reaches a level of around 300 W after 6 min. Taking this heat source away from the engine results in a negative temperature offset of around 1.5 K. A significant deterioration of fuel consumption could not be measured. Significant influence on exhaust gas temperature and mass flow could not be found in the measurements. Any impact on the WHR-system will be discussed in the following.

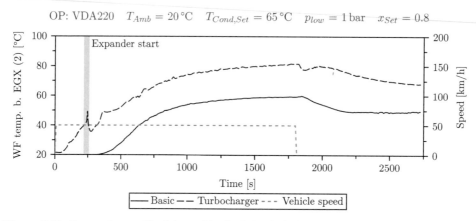

Figure 5.21: Temperature profile of the working fluid ahead of exhaust gas heat exchanger (2) for system configurations with and without turbocharger waste heat use

The effect of the addition of the waste heat from the turbocharger is depicted in figure 5.21. By using the turbocharger heat it is possible to increase the temperature before the EGX earlier than the basic configuration and it is also possible to reach a higher level. At the end of the 50 km/h part, the temperature reaches 82°C which is higher than the level of the recuperator (68°C), but smaller than the level of the preheater(93°C). A combination of turbocharger heat and recuperator showed an identical behavior in the beginning, but reached a temperature level of 88°C in the end.

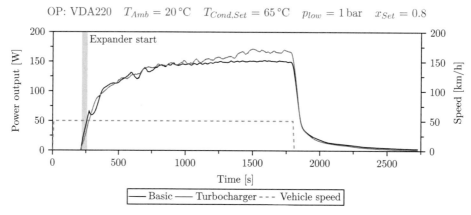

Figure 5.22: Power output for system configurations with and without turbocharger waste heat use (floating average over 30 s)

As all previous results showed, improving the temperature ahead of the EGX increases power output of the system. A comparison between the addition of the turbocharger heat and the basic measurement is given in depiction 5.19. The power output is enhanced in the steady-state phase by the use of the turbocharger heat without the start-up and the heat-up phase being altered. This can be explained in the same way as for the preheater and the recuperator. Yet using the turbocharger waste heat proofs to be beneficial in cold start.

5.1.8 Conclusion

The system behavior in cold start was investigated in this chapter. It was found that the system heats up one component after the other. A phenomenological model was given wherein the heat wanders through the system like a wave and heats up the EGX first, then the expander and after that the condenser and pump. Only when the temperature before the EGX reaches its operating temperature, the system has reached steady-state. The heat-up of the expander proofed itself to be the most important aspect for the system in cold start to improve power output.

The start / ambient temperature was found to have no measurable influence on the systems power output. The cold start of the engine and the resulting increase in fuel consumption compensate the additional heat demand.

Steam quality and mass flow were investigated for benefits in heating up the system. Yet only small potential was found due to the chronological need to set the optimal steam quality at the time of the expander start-up.

Low-temperature and low-pressure showed the same influence as in steady-state operation. The low-temperature should be as high as possible and the low-pressure as close to the optimal pressure ratio as possible.

The addition of further heat sources in the form of a recuperator, a preheater or a turbocharger displayed improvements in the working fluid temperature before the EGX and the final temperature level. Yet they were not as beneficial in the first phase of the cold start, as they did not improve the heat-up process of the expander immensely. The power output in the steady-state section of the driving cycle on the other hand did benefit from the additional heat sources.

5.2 Synergies in the use of Condenser Heat

Except for the recovered energy that can be used to improve the fuel economy of the car, there is another possible benefit from a waste heat recovery system. The direct use of the exhaust heat is beneficial for the car, as explained in chapter 2.2.1. However, it is also possible to use the remaining heat of the WHR-system to improve the cold start behavior of the engine and the passenger cabin. An improved warmup of the engine has benefits in terms of fuel consumption and exhaust emissions [45],[34],[6]. An improvement resulting from the use of the WHR-system can be achieved by connecting the condenser to the cooling circuit of the engine. Most of the heat flow of the EGX is not transformed into mechanical / electrical energy by the expander but is dissipated by the condenser. This heat could be used to increase the temperature of different parts of the engine and the cabin. Through the double heat exchange over the EGX and the condenser and also through the expansion process itself, the remaining temperature level that can be achieved is most likely lower than the operating temperature of most components. This temperature level depends on the low-pressure and low-temperature-level and on the flow of coolant through the condenser. As low-temperature and low-pressure have their optimal settings in terms of WHR-system power output, the volume flow was found to be the lever to increase the coolant temperature after the condenser. To find the optimal settings and to investigate the potential, several tests were performed that are presented in the following. Parts of these results were also presented in [76].

The tests in this chapter were performed with active passenger cabin heating. The cabin heating was reproduced by a plate heat exchanger that was controlled to have the desired heat flow withdrawal from the engine coolant. This desired heat flow profile was adapted from a simulated cold start[3] run with a car model of the same middle class passenger car. The heat flow profile was kept the same for all investigations. All other settings were identical to the ones from chapter 5.1. The first result is the heat flow from the condenser to the coolant in a base configuration. The behavior is depicted in figure 5.23.

Config.: Basic   OP: VDA220 $T_{Amb} = 20\,°C$ $T_{Cond,Set} = 65\,°C$ $p_{low} = 1\,\text{bar}$ $x_{Set} = 0.8$

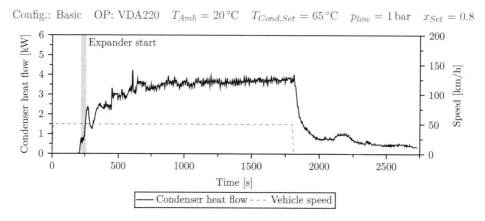

Figure 5.23: Condenser heat flow for a VDA220 starting at 20°C

For the given settings the condenser is able to deliver a heat flow of around 3.7 kW. This heat can be supplied after ca. 870 s. The first heat can be drawn from the condenser after about 200 s, which is only seconds after the expander started working. Again the wave-like behavior of heating up one component after another is the reason. The power output of the system as well as the condenser heat flow are almost independent from the starting temperature due to the characteristics of a SI engine in cold start (catalyst heating, increased engine load through friction).

5.2.1 Influence of Condenser Coolant Flow

As mentioned before, the coolant temperature downstream of the condenser mostly depends on the coolant mass (volume) flow for most operating points. Tests with coolant flows at 2, 5 and 10 lpm were performed to find which coolant temperatures would result after passing the condenser with an inlet temperature of 65°C. The condenser circuit was separated from the engine for these experiments, just as for the earlier experiments. The outcome is presented in figure 5.24.

The temperature of the condenser coolant could be increased from 65°C to around 71°C for 10 lpm, to 77°C for 5 lpm and around 92°C for 2 lpm. The 2 lpm volume flow produced the highest outlet temperatures, and would have been chosen for the following measurements, if it wasn't for some sever problems with the heat rejection due to local boiling within the condenser. If the volume flow was any lower, it could happen that the working fluid would not be cooled down far enough which would result in cavitation and a collapse of the working fluid mass flow. To prevent that from happening a volume flow rate of 2.5 lpm was chosen, if not mentioned otherwise. To prevent unnecessary heat losses the flow was prohibited until the condenser needed cooling.

---

[3]The simulation results were provided by a cooperation partner.

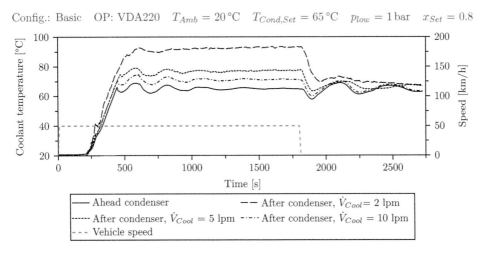

Config.: Basic   OP: VDA220 $T_{Amb} = 20\,°C$ $T_{Cond,Set} = 65\,°C$ $p_{low} = 1\,bar$ $x_{Set} = 0.8$

Figure 5.24: Coolant temperature profile after the condenser (1) for different volume flow rates

Lowering the condenser coolant volume flow is only reasonable, if the heat flow that needs to be rejected from the working fluid can be maintained. Otherwise the working fluid would not be cooled down far enough (as it seems to be for the 2 lpm volume flow) and the rejected heat that could be utilized would be lowered. In order to investigate this influence, simulations and measurements were performed. At first the heat flows were calculated for the before mentioned measurements and a simulation was performed for further volume flow rates and to validate the measurement results. The findings are depicted in figure 5.25 and 5.26.

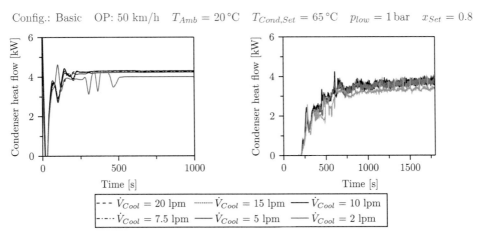

Config.: Basic   OP: 50 km/h $T_{Amb} = 20\,°C$ $T_{Cond,Set} = 65\,°C$ $p_{low} = 1\,bar$ $x_{Set} = 0.8$

Figure 5.25: Condenser heat flow for different volume flow rates - simulation results for 50km/h OP

Figure 5.26: Condenser heat flow for different volume flow rates - measurement results for 50km/h OP

Figure 5.25 shows that the condenser heat flow is practically independent from the coolant volume flow as long as a certain threshold is not undercut. For the 50 km/h OP this limit seems to be around 2 lpm. The heat flow is slightly lowered and the simulation starts to run into instabilities. The reason for this independence can be found in the dimensioning of the condenser which is designed to be able to transfer heat flows for much more demanding OPs (> 120 km/h). The heat exchanging surface is large compared to the heat flow. Thus the heat is transferred anyway as long as local boiling is prevented (for example by higher coolant pressures) and as long as a minimum volume flow is supplied. Figure 5.26 proofs that this trend can be also seen in real measurements. The overall heat flow is slightly smaller due to additional heat losses by radiation and convection to the environment.

The results show that the condenser coolant volume flow should be as small as possible and within the security range in order to achieve high coolant temperatures at the outlet. Thus the location where the coolant is taken from the coolant circuit of the car and the location where it is fed back have to be chosen wisely to gain possible benefits from the condenser heat.

5.2.2 Influence of Coolant Circuit Connection

With the knowledge of the basic behavior from the previous investigations, different coolant circuit configurations were tested. These were each designed to improve the heat-up behavior of the engine with each configuration having its focus one main component. These components were: the cabin heater (V2), the engine (V3), the gearbox (V4), a second cabin heater (V5), the engine oil cooler (V6) and a basic configuration with separated coolant cycles (V1). Detailed explanations and depictions of each configuration can be found in appendix A.6. The temperatures of engine coolant, engine oil, gearbox oil and engine oil, as well as fuel consumptions were chosen to represent the benefits in warm-up. They will be discussed in the following.

Config.: Basic   OP: VDA220 $T_{Amb} = 0\,°C$ $T_{Cond,Set} = 65\,°C$ $p_{low} = 1\,bar$ $x_{Set} = 0.8$

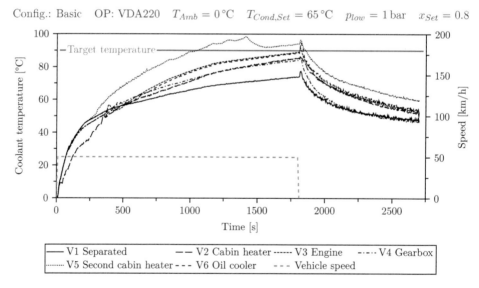

Figure 5.27: Coolant temperatures in the cylinder head for a VDA220 starting at 0°C

The engine coolant temperature was measured within the cylinder head to receive the core temperature of the engine. This temperature is relevant for the wall heat transfer between the coolant and the gas inside the cylinder. The hotter it is, the less heat is lost to the coolant and the less fuel needs to be burned. All configurations lead to an increase of the temperature at the end of the 50 km/h part (1800 s) due to the additional heat from the condenser, as can be observed in figure 5.27. V5 was even able to trigger the thermostat to open. V2 is the only configuration that was disadvantageous compared to the standard V1. This was the case for the first minutes and was due to its 5 lpm coolant flow from engine start on, which transported cold water from the condenser into the engine.

The engine oil temperatures were found to behave similar to the coolant temperature, thus they are not presented separately. All configurations were able to increase the temperature with V5 having the highest influence and V6 the second highest. The connection between coolant temperature and oil temperature can easily be explained through the exchange of heat within the engine oil heat exchanger.

Config.: Basic   OP: VDA220 $T_{Amb} = 0\,°C$ $T_{Cond,Set} = 65\,°C$ $p_{low} = 1\,bar$ $x_{Set} = 0.8$

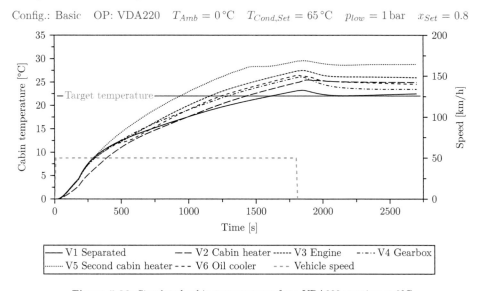

Figure 5.28: Simulated cabin temperatures for a VDA220 starting at 0°C

Figure 5.28 shows the simulation results of the cabin temperature for the different configurations. The results were calculated for a middle class passenger car. Even though the heat flow over the cabin heat exchanger was kept constant for all measurements, each configuration resulted in a different simulated cabin temperature. This was due to the differences in coolant temperature of the coolant that flows through the cabin heater and the resulting higher air flow temperature level. To keep the cabin heater heat flow constant in the simulation, the air flow was adapted to fit. The differences in cabin temperature thus result from the higher heating potential of the hotter air that is blown into the cabin.

All configurations achieve temperature levels higher than the target temperature in the end. The reason is that the heat flow was not restricted in the measurements when the cabin would potentially have reached the target. The aim was to observe the differences until 22°C are reached. The standard configuration needs 1600 s to reach this level. With a second cabin heater (V5) the time could be shortened to 940 s. In the other configurations it took 1160 s (V3) to 1430 s (V2).

These results are supposed to give a comparative view on the different coolant circuit configurations, but are not supposed to represent precise outcomes for the achievable cabin temperature levels. The variants with higher coolant temperatures would also have been able to transfer higher heat flows into the cabin thanks to their higher temperature level. But that would have decreased the improvement in engine warm-up. Real tests with a real passenger car on a climate wind tunnel would need to be performed in order to receive precise results. Yet theses simulations show that a WHR-system is able to improve the heat-up of a passenger cabin if it is connected to the coolant circuit in the suitable way.

Config.: Basic   OP: VDA220 $T_{Amb} = 0\,°C$ $T_{Cond,Set} = 65\,°C$ $p_{low} = 1\,bar$ $x_{Set} = 0.8$

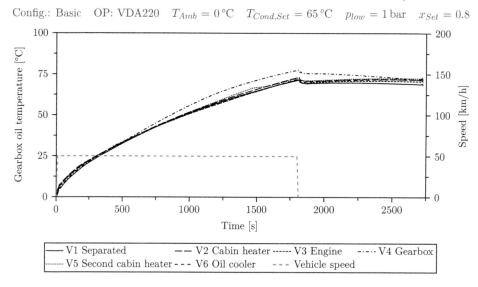

Figure 5.29: Gearbox oil temperatures for a VDA220 starting at 0°C.

The gearbox oil temperature is an indicator for the friction losses within the gearbox. Cold oil has a higher viscosity and causes higher friction. The resulting loss in torque has to be compensated by higher engine load and thus higher fuel consumption. Hepke predicted a fuel consumption improvement of 2.1 % in a NEDC with improved gearbox thermalmanagement [45]. The measurement outcomes for the coolant circuit configurations are depicted in figure 5.29. The influence of most configurations on the temperature is not significant. Only V4 is able to raise it, yet the advantage sets in late. At this point in time the oil had already reached 35°C and the viscosity was significantly lowered. Overall the possible impact on warm-up is low.

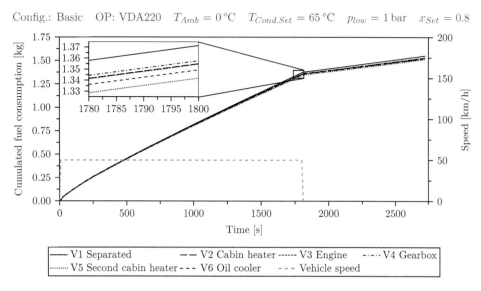

Figure 5.30: Cumulated fuel consumption for a VDA220 starting at 0°C

Based on the advantages in engine coolant temperature and gearbox oil temperature the fuel consumption could be reduced for all configurations. Figure 5.30 shows the cumulated mass flow over time. In order to make the differences between the configurations more perceptible, the final section of the 50 km/h driving cycle part was enlarged. As shown, all configurations improved the fuel consumption. V5 showed the smallest fuel demand due to its highest engine coolant and oil temperature levels. V4 had no advantages compared to the other configurations, which can be explained through the late set in of its benefits.

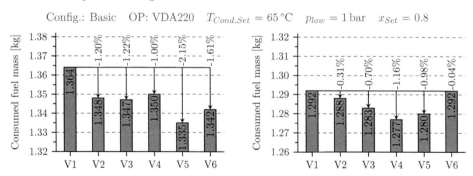

Figure 5.31: Consumed fuel after 1790s of a VDA220 cycle at 0°. No power feedback from WHR-system

Figure 5.32: Consumed fuel after 1790s of a VDA220 cycle at 20°. No power feedback from WHR-system

The fuel consumption could not only be lowered for starting temperatures of 0°C but also for starting temperatures of 20°C as figure 5.32 shows. The advantage is smaller at higher starting temperatures which is caused by the lowered potential to reduce the friction mean effective pressure [36]. The reduction in these cases is only due to improved thermal management. Any benefits based on the feedback of electric power from the WHR-system would come on top of these.

For 0°C fuel consumption could be decreased between 1-2 % in the VDA220. For 20°C the improvement was around 1 %. V5 was still the one with the second lowest fuel consumption. V4 seemed to have a bigger impact on fuel consumption, but the cabin heater did not consume the correct amount of heat from the system in this test run due to a measurement error.

5.2.3 Conclusion

This chapter investigated possible synergies that can be gained by purposeful use of the remaining condenser heat. The basics of the condenser heat flow were covered by testing different coolant volume flows in real measurements and in simulation. These results show, that it is possible to clearly increase the outlet temperature of the coolant by reducing the volume flow. The heat flow can be kept constant if the volume flow is above a certain limit to prevent local boiling of the coolant. Otherwise the heat flow is reduced and the working fluid is not cooled enough resulting in a collapse of its mass flow.

Six different coolant circuit configurations were conceived, that were realizable and possibly beneficial. Each variant is discussed and depicted. All configurations were tested on the test bench at 0°C and 20°C. The use of condenser heat proved to be beneficial in all configurations. The engine warm up can be accelerated and thus fuel consumption can be reduced, while also increasing comfort through higher cabin temperatures. However, considering the right volume flow rates and the right choice of coolant circuit connection is important to gain these improvements. The conclusion is that WHR-systems can reduce fuel consumption not only through power generation, but also through improved warm-up.

6 Summary and Conclusion

"Increasing the resource efficiency represents one of the key challenges of the 21st century" [81]. This statement published by the Verein Deutscher Ingenieure (VDI) is the driving force behind the work of most engineers in the field of thermodynamics. In the area of passenger cars this issue is mostly present in the form of fuel consumption. As combustion engines are and still will be the most common and certainly most resource-conserving way of propulsion, it is an essential task to steadily improve them.

The technological path of waste heat recovery is a new topic for passenger and commercial vehicles. One of the most promising technologies in this field is the Clausius Rankine cycle. The exhaust gas heat flow is used to vaporize a working fluid which propels an expander and thereby delivers usable energy. Even though the basic principle has been used in advanced forms in power plants for more than 110 years [101], the use in passenger cars is a very recent approach. The boundary conditions demand concessions in terms of complexity, size, weight and operating parameters.

As mentioned in the introduction most of the work that has been previously done in this field is of theoretical nature or investigates single components under laboratory conditions. Only a small number of studies with real and viable hardware has been published. This work shall be one of them. The results and statements are based on the measurement results gained on a real WHR-Rankine-system that was added to a real, state-of-the-art gasoline DI-engine. All components are as close to production level as currently possible. A basic system was created solely for this purpose that was designed in earlier projects. The main components were a gear type pump, a plate exhaust gas heat exchanger, a scroll expander and a plate heat exchanger as a condenser. The engine combined with the WHR-system was put on a temperature-controlled engine test bench, allowing to investigate the system behavior under a wide range of boundary conditions.

6.1 Summary

This work aims at contributing to the topic by investigating the possibilities of waste heat recovery systems to improve fuel consumption of passenger cars. In order to do so, several experimental studies and simulations were performed. The experiments encompass steady-state operation, the possibilities of additional heat sources, the influence of engine parameters and the system behavior during cold start. Additional simulations were performed when reasonable and adding information to the experimental findings.

The first task was to characterize the fundamental influence of system operating parameters. The most relevant parameters that can be changed during operation are the mass flow of working fluid which is dictated by the rotational speed of the pump, the low-temperature which is directly linked to the condenser coolant temperature and the low-pressure that is given by the pressure within the expansion reservoir. These parameters were compared based on their resulting heat flows and mainly on their impact on the power output. The results were mostly presented for a 50 km/h operating point of a mid-sized car engine.

The mass flow was found to have an immense influence on the power output of the system. The steam quality can be directly controlled by setting a desired mass flow. The higher the steam quality the higher the power output. Due to the limitation of the expander, which needed wet steam, only steam qualities up to 0.9 could be investigated. A quality of 0.8 was chosen for the tests as a safety measure. The condenser coolant temperature also had a significant influence on the system. Higher condenser coolant temperatures reduce the heat loss over the condenser and thus increase the temperature before the exhaust gas heat exchanger. For a given low-pressure the condenser coolant temperature should be as high as possible to enhance the power output. The low-pressure should be adapted to the current heat flow of the EGX. In principle the low-pressure should be as low as possible, yet it is restricted by the risk of boiling ahead of the pump and by the pressure ratio of the expander. For small amounts of heat input the low-pressure should be set in relation to the high pressure to optimally meet the pressure ratio of the expander. When the heat input climbs, the high-pressure climbs with it. In these cases the low-pressure would need to be set to higher values. A tradeoff between the optimal pressure ratio and low low-pressures occurs. The low-pressure should thus be adaptable to the current demand.

The requirement of simple system design can be adjusted if an alteration justifies the additional complexity. This was tested for several system configurations that had additional heat inputs, including a partial flow recuperator, a preheater, a full flow recuperator and the integration of the turbocharger coolant channels into the system. Each additional heat source was able to enhance the temperature of the working fluid ahead of the EGX which improved the power output. The enhancement formed depending on the temperature level of the source and its amount of deliverable heat. The heat sources also differ in their influence on the condenser heat flow. While the recuperator took heat away, the other sources increased the heat flow. The choice of an additional component depends on the additional cost and the possible impact on the CO_2 emission and should be assessed depending on the engine.

Engine operating parameters were investigated to present information about the basic interaction between the engine and the waste heat recovery system and also to search for possible setups that improve the overall system efficiency. This could be either achieved by keeping the efficiency of the engine constant and increasing the power output of the WHR-system or by lowering the engine efficiency and overcompensating the loss with the WHR-system. Additionally, the emission levels of pollutants should be unfazed by the alteration. Unfortunately, no parameter could be found that would increase the overall efficiency. The efficiency of the combustion process is always more relevant to the overall efficiency than the one of the WHR-system. Yet the ignition timing was found to be a powerful lever on the power output of the WHR-system. By setting a late ignition timing it is possible to increase the output by more than three times, which provides a possible lever for increasing the power output of the WHR-system if desired.

With the behavior in steady-state investigated thoroughly, it was possible to investigate the system in cold start scenarios. The VDA220 driving cycle covering ambient temperature from -10°C to +20°C was chosen for this purpose. The results for cold start operation presented in this work are the first published results for the application of a Rankine-system on a combustion engine. The results showed that power is delivered a few minutes after the engine is started, but it takes between 20-30 min until the system reaches absolute steady-state conditions. The expander starts turning just after it is entirely filled with steam. The timing as well as output level seem to be independent from the ambient temperature. This can be attributed to the increased fuel consumption of the engine at lower starting temperatures. Changing the system operating parameters did not seem beneficial for cold start. Additional heat sources were only able to improve the power output just before steady-state was almost reached. Yet an improvement was reached through their implementation.

Possible synergies through the use of condenser heat were investigated as a last facet of the overall system approach. The remaining heat that the condenser transfers to the coolant circuit can be used to improve the warm-up of different parts of the car. The results show that a Rankine-WHR-system is not only able to reduce fuel consumption through producing mechanical or electrical power, but also through regaining heat in cold start operation. The warm up time of the engine and its components as well as of the passenger cabin can be reduced. This decreases the fuel consumption and enhances the driver's comfort and safety. Different configurations for the coolant system connection to the condenser are possible with each offering specific benefits. Especially a configuration using a second cabin heater appears very efficient. With the combination of generated power and improved warm up, Rankine based WHR-systems are able to reduce CO_2 emissions for future combustion engines.

6.2 Evaluation

The application of waste heat recovery systems in passenger cars offers a way to improve fuel consumption in the future. It is possible to reduce the fuel consumption in real driving conditions, but the technology needs further improvements to reach the level of possible deployment. The systems need to be absolutely reliable to be accepted by the consumers. A faulty system might lead to a stalemate of the whole car either through failure in the cooling system, the bypass or due to regulatory intervention caused by the not working CO_2 reduction measure. This must be avoided at all costs. A lot of time and effort needs to be put into the development of components to provide this level of security. The exhaust gas heat exchanger needs to be able to withstand numerous thermal cycles without leaking, the pump must be able to provide high pressures over its lifetime and the expander needs to be resistant to wear.

Any improvement of the combustion efficiency of an engine is most likely to have a negative impact on the power output of the WHR-system. As the results from chapter 4.3 show, this can be considered positive for the overall efficiency of the car, but lowers the benefit of a WHR-system. Alternative combustion processes like homogeneous charge compression ignition or stratified operation also stress this benefit as they improve engine efficiency and in the case of stratified operation also increase the air flow, which reduces the exhaust gas temperature [56].

Cold start of systems constitutes another challenge. In order to achieve significant improvements in official driving cycles, the systems need to be able to quickly reach their operating temperature. The tested system needed four minutes to deliver the first output, which is considered to be quite late considering the duration of the European driving cycles[1]. Additional heat sources for the working fluid did not improve this timing. Other solutions need to be found.

At the moment this level of sophistication is not reached and it does not seem to be reachable in the next years as most car manufacturers have to focus their efforts on electric vehicles and the fulfillment of pollutant emission levels. But when the interest in this technology is rekindled and the needed degree of maturity is reached and systems are being deployed in passenger cars, they will help increase the resource efficiency of combustion engines by reducing the fuel consumption

---

[1]NEDC about 20 min, WLTC about 30 min

A Appendix

A.1 Working principle of a scroll expander

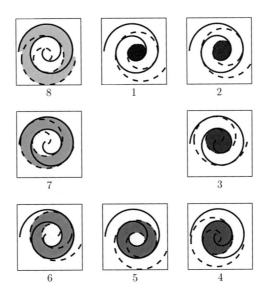

Figure A.1: Representation of the working principle of a scroll expander by Körner [57]. The rotating spiral is represented by the dashed line and the fixed spiral by the solid line. The encased volume is represented by the gray area.

A scroll expander consists of two intertwined spirals (scrolls) of which one is stationary and one can rotate on a circular path. The separated chambers enlarge during the rotation and thereby expand the steam. The expansion propels the rotating scroll which is connected to an eccentric shaft, where the generated power can be withdrawn. The enclosed volume is separated by a seal on the front side of the spiral and by fluid liquid at the line where both spirals come closest together (but never touch).

A.2 Depictions of relevant Thermodynamic Cycles

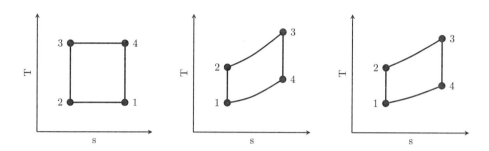

Figure A.2: Schematic T-s-Diagram of a Carnot cycle, a Otto cycle and a Brayton cycle

These cycles represent the most relevant thermodynamic cycles that are explained in chapter 2.1.5.

A.3 Schematic of a preheater supplied with engine coolant

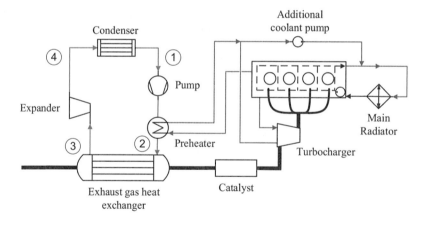

Figure A.3: Schematic of the advanced system consisting of the WHR-system including a coolant (heated by the engine) preheater, the exhaust gas system and the engine cooling system

The measurement results from chapter 4.2.2 were gained with a preheater that used an electric heat source to keep the engine unaltered. However, using the coolant circuit of the car as source would have been possible nonetheless. A possible configuration is presented in figure A.3. The coolant should be withdrawn from a location with maximal temperature level, e.g. the outlet of the cylinder head. Temperatures of around 105°C may be reached and were tested on the test bench. After the coolant has been used by the preheater it can be redirected to the coolant circuit. It is either reasonable to redirect it before the main radiator and thus have an optimal heat discharge or redirect it to the coolant pump and thus drop the additional pump.

A.4 Simulation Models

Combustion Engine Model

The engine model was used to extend the measurement results in chapter 4.3 by generating values for exhaust gas temperature and mass flow that could be used by the Rankine-system model. It was a regular model of a four cylinder turbocharged engine that was adjusted to fit to the engine used for the experiments. The geometries of piston, cylinder and valves were changed. Ignition, injection and valve timing were adjusted to their real values. To keep complexity and calculation time manageable the coolant and oil circuits of the engine were not simulated. Intake system, valve train and exhaust system were constructed with great detail to give the desired precision. The model uses a GT-Suite internal turbulent flame model to predict the in-cylinder burn rate for spark-ignited engines with homogeneous fuel-to-air mixtures. Geometries and valve timing are used to estimate the in-cylinder flow and turbulence. A vaporization and an injection model are used to calculate local lambda values. In combination with the ignition location it predicts a flame propagation. This way a heat release rate can be predicted. The wall heat flow is simulated using a fitted Woschni approach. [31]

The whole model was validated for the operating points 50 km/h and 120 km/h. Further details can be found in the work of Lagaly [61].

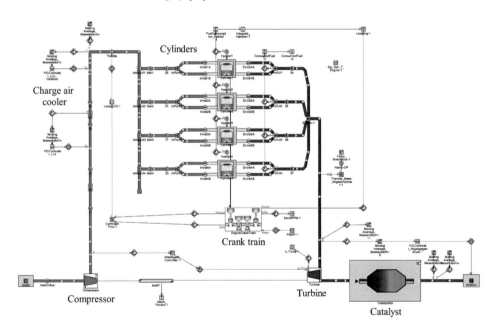

Figure A.4: Simulation model of the investigated combustion engine in GT-Suite

Rankine-System Model

Figure A.5 is a depiction of the WHR-system simulation model in GT-Suite. The model consists of the main parts pump, exhaust gas heat exchanger, expander, condenser and expansion reservoir. Similar model were created which included heat exchanger which represented the functions of recuperator or preheater.

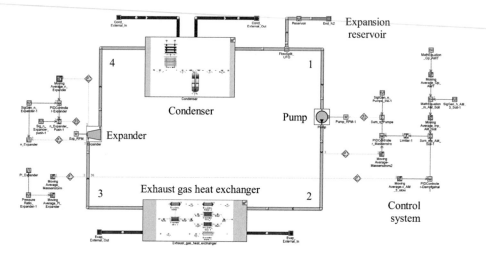

Figure A.5: Simulation model of the basic WHR-system in GT-Suite

Passenger Cabin Model

The cabin model consists of a cabin heat exchanger (or two for the investigations in chapter 5.2.2) and a passenger cabin template. The model takes a given environment air mass flow and uses the engine coolant to heat it up. The hot air is then used to increase the cabin temperature. Inputs are mass flow and temperature of air and coolant aswell as ambient temperature and vehicle speed. The cabin template takes multiple boundary conditions into account. The size of the cabin, materials and thermal mass of different parts of the cabin, solar flux, radiation, convection and humidity are examples of the main influences. It was not possible to compare the results to real measurements and thus the absolute values should be acknowledged to have an uncertainty, but the tendencies should be reliable.

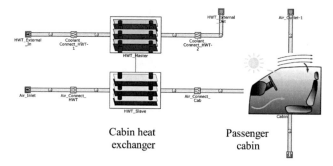

Figure A.6: Simulation model of the passenger cabin of a mid-size car in GT-Suite

A.5 Influence of steam quality for different working fluids

The influence of low-temperature and low-pressure for different working fluids was investigated in chapters 4.1.2 and 4.1.3. The following two figures display that the same statements apply to other steam qualities and different working fluids. The results bear some uncertainty as the model they were calculated with was calibrated for steam qualities of 0.8, nonetheless the tendencies are clearly visible.

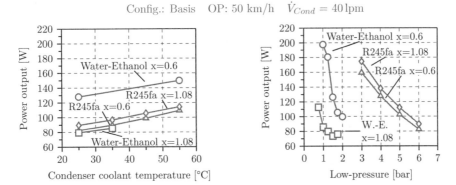

Figure A.7: Simulation results for the influence of steam quality on power output for different working fluids over condenser coolant temperature. Low-pressure is 1.5 bar for the water-ethanol-mixture and 6 bar for R245fa.

Figure A.8: Simulation results for the influence of steam quality on power output for different working fluids over low-pressures. Condenser coolant temperature is 25°C.

A.6 Possible Coolant System Connections

Several different coolant circuits were conceived in order to use the remaining heat from the condenser to improve the warm-up of different parts of the engine. Not all possible configurations are reasonable. The condenser's coolant entrance temperature is limited by the requirement to liquefy and subcool the working fluid. Furthermore the coolant's temperature at the condenser outlet has to be higher than that of the component to be heated. Therefore the possibilities are limited. The investigated configurations will be explained in the following paragraphs.

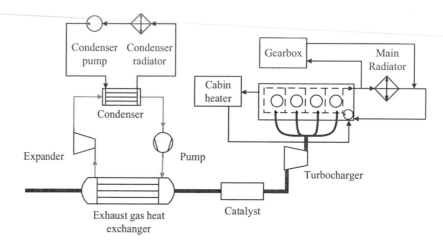

Figure A.9: Schematic of the cooling systems of WHR-system and engine - V1 separated cooling circuits. The flow directions of EGX and condenser are not depicted in the physically correct way (countercurrent) from here on, due to reasons of graphical representation

(V1 Separated) In the basic configuration the coolant circuits are separated. The circuits don't interact and the engine is not influenced, as there is no additional heat flow to the coolant circuit. The coolant pump of the engine supplies the gearbox, the cabin heater and the main radiator with coolant as soon as the thermostat opens. The condenser has its own pump and radiator. The coolant temperature ahead of the condenser is limited to 65°C and the volume flow rate is set to 2.5 lpm, which is the case for all configuration (except mentioned otherwise).

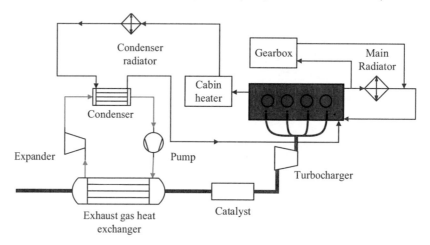

Figure A.10: Schematic of the cooling systems of WHR-system and engine - V2 reheating the coolant after the cabin heater

(V2 Cabin Heater) The next possible configuration is to divert coolant after passing the cabin heater and reheat it with the condenser before transporting it to the engine. In order to secure proper cabin heater operation, the volume flow through the cabin heater is set to a higher value of 5 lpm which decreases the outlet temperature of the condenser but also allows omitting the condenser pump. The condenser radiator must be kept for operating points without cabin heating demand.

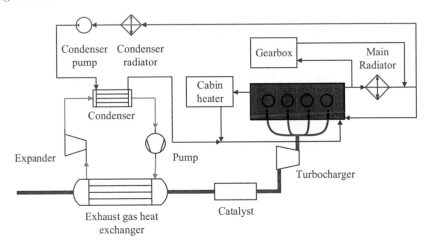

Figure A.11: Schematic of the cooling systems of WHR-system and engine - V3 heating the coolant after the main radiator

(V3 Engine) Another configuration is to withdraw cold coolant downstream of the main radiator and then recirculate the heated coolant ahead of the engine coolant pump. The condenser radiator is kept to ensure that the coolant temperature ahead of the condenser stays below the defined limit if needed.

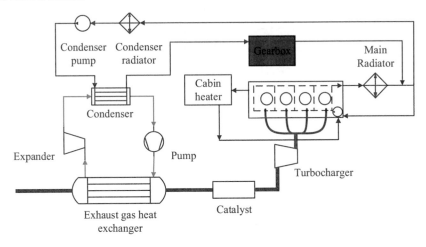

Figure A.12: Schematic of the cooling systems of WHR-system and engine - V4 using the condenser heat to heat up the gearbox

(V4 Gearbox) To improve the warm up of the gearbox a configuration can be chosen, where coolant is withdrawn after passing the main radiator and is heated by the condenser before entering the gearbox oil cooler. Thus the gearbox can be heated without interfering with the engines coolant flow. The aim is to decrease oil viscosity and thus friction. This is predicted to be the most beneficial target for additional heating in the works of Hepke [45] and Lee [64]. The condenser pump and radiator have to be kept for post-heat-up operation.

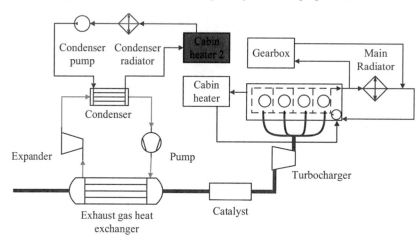

Figure A.13: Schematic of the cooling systems of WHR-system and engine - V5 extra cabin heater

(V5 Second Cabin Heater) Another way to use the condenser heat flow is to apply a second cabin heater. For larger passenger cars this is often already the case. In this case the second cabin heater is only connected to the condenser and can run at full volume flow of 10 lpm. The overall heat flow to the cabin was kept constant by reducing the heat flow of the first cabin heater. Therefor results can still be compared.

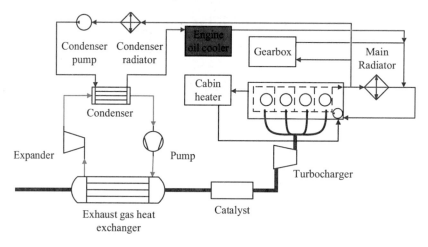

Figure A.14: Schematic of the cooling systems of WHR-system and engine - V6 using the condenser heat to heat up the engine oil

(V6 Oil Cooler) The last conceived configuration is similar to V4, but instead of the gearbox, the engine oil cooler is heated by the condenser.

A.7 Diesel Combustion

A short investigation into the possibilities and coherences between the engine operating parameters and the exhaust gas enthalpy flow of a diesel engine was performed analogous to the investigations on the gasoline engine used prior. Diesel combustion is commonly considered to have less potential in waste heat recovery as its exhaust gas heat flow features lower temperature levels compared to spark-ignition combustion (Otto) [111].

The engine was a 2.0 l common rail diesel engine with an exhaust gas turbocharger and high-pressure (exhaust gas is withdrawn ahead of the turbocharger) exhaust gas recirculation (EGR). The engine was also equipped with thermocouples in the exhaust pipe. The mass flow of exhaust gas was calculated using the measured air and fuel mass flow. As the influences of the center of the combustion and most other operating parameters were expected to be similar, the investigations concentrated on the influence of the EGR rate. This value was altered by the ECU by changing the alignment of the EGR-valve. The "real" EGR-rate could not be measured on this setup. Yet the position of the valve gives a clue on the amount: a more opened valve results in higher EGR-rates. No WHR-system was installed on the engine so the influence on power output could only be determined using the simulation model. The investigations were performed for the 50 km/h operating point.

OP: 50 km/h   Engine: Diesel

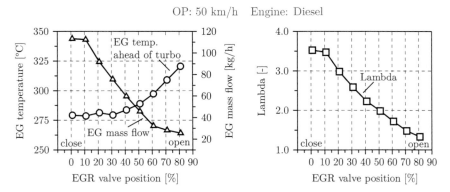

Figure A.15: Influence of EGR valve position on exhaust gas temperature and mass flow

Figure A.16: Influence of EGR valve position on lambda values

Figure A.15 shows the behavior of exhaust gas temperature ahead of the turbocharger and the mass flow of exhaust gas. Mass flow and temperature behave contrary to each other. Higher EGR-rates lead to higher temperatures but smaller mass flows. The EGR is substituting intake air, but at the same time lowers the amount of fresh air that can be delivered to the engine. The added heat from the combustion is distributed over a smaller amount of charge, which leads to higher exhaust gas temperatures. Compared to the gasoline engine the temperature levels are clearly lower even at higher EGR-rates; the mass flow is similar for small EGR-rates but considerably smaller at higher EGR-rates. The resulting change in exhaust gas composition can be observed in figure A.16 where the lambda value is depicted. The recirculated EG substitutes parts of the intake air thus the amount of oxygen in the exhaust gas is reduced. This effect was not considered in the simulation model as it was created for stoichiometric combustion, but the influence was considered minor.

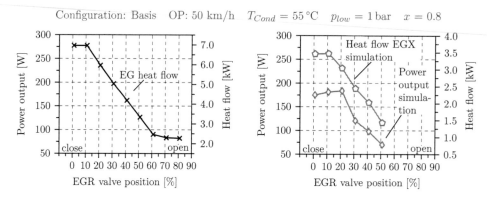

Figure A.17: Influence of EGR valve position on EG heat flow

Figure A.18: Simulation results for the influence of EGR valve position on EGX heat flow and power output

The calculated exhaust gas heat flow depicted in figure A.17 shows that the result of the change of EGR-rate is a reduction of heat flow. The increase in temperature does not compensate the reduction in mass flow. The investigations of Jacobs showed similar results in terms of exhaust characteristics, but lack a simulation of the power output of a potential system [49]. The simulations (only calculated for rates up to 50 %) show a similar behavior in the EGX heat flow which can be observed in figure A.18. The power output is also reduced by the smaller EG heat flow. The conclusion is that for the purpose of a WHR-system the EGR-rate should be as small as possible as the power output could be similar (a comparison with a real system should be performed to validate this) yet with the need of higher EGR-rates to reduce emission levels the power output will be reduced.

A.8 Measurement uncertainty

Every measurement is prone to a certain amount of inaccuracy and as calculated values (like heat flow, steam quality etc.) depend on measured values that also include uncertainty. An overview of the measurement inaccuracy of the most important utilized devices is listed in table A.1.

Table A.1: Measurement uncertainty of the utilized measuring devices

Parameter	Principle	Range	Uncertainty
Mass Flow	Coriolis	0-260 kg/h	± 0.1 % of reading
Pressure (High)	Piezo-resistive	0-25 bar	± 0.05 % of full scale
Pressure (Low)	Piezo-resistive	0-4 bar	± 0.05 % of full scale
Temperature (Engine)	Thermocouple (Type K, Class 1)	-40 to 1000°C	1.5 K or 0.4 % of reading
Temperature (WHR-System)	Resistance (PT100, Class A)	-50 to 560°C	0.35 - 1.05 K depending on temp.
Torque	Strain gauge	± 1000Nm	<± 0.05 % of full scale
Volume Flow	Turbine frequency	0-10/25/50/70 lpm	± 0.5 % of reading

Other uncertainties arise from unmeasurable values and stochastic variations. These include the amount of oil that is cycling through the system at a given point in time. The amount of heat that is transferred by the EGX in relation to the amount that is lost to the environment. The real mass ratio of the working fluid. An assessment of the magnitude of these failures can be found in the work of Körner [57]. He calculated the measurement uncertainty according to Gauß's propagation of uncertainty method for exemplary values of enthalpy, heat flow, power output and efficiency for a WHR-system.

Bibliography

[1] ARIAS, Diego ; SHEDD, Timothy ; JESTER, Ryan: *Theoretical Analysis of Waste Heat Recovery from an Internal Combustion Engine in a Hybrid Vehicle.* Bd. 2006-01-1605. SAE, 2006

[2] ARMSTEAD, John R. ; MIERS, Scott A.: *Review of Waste Heat Recovery Mechanisms for Internal Combustion Engines.* ASME Journal of Thermal Science and Engineering Applications, 2014

[3] BAEHR, Hans: *Wärme- und Stoffübertragung.* 9., aktual. Aufl. 2016. 2016 (SpringerLink : Bücher). – ISBN 3662496763

[4] BAEHR, Hans ; KABELAC, Stephan: *Thermodynamik: Grundlagen Und Technische Anwendungen.* Vieweg + Teubner Verlag, 2017. – ISBN 3662495678

[5] BAUER, Felicitas: *Theoretische Untersuchung zur Verdampfungsinstabilität in einem Plattenwärmeübertrager mit Strömungskanälen.* Bachelorarbeit. KIT, 2016

[6] BLODIG, Stefan: *Warmlauf des Verbrennungsmotors im Hybridfahrzeug.* Dissertation. Technische Universität München, 2010

[7] BÖCKH, Peter ; WETZEL, Thomas: *Wärmeübertragung: Grundlagen und Praxis.* 7., aktualisierte u. überarb. Auflage 2017. 2017. – ISBN 978-3-662-55479-1

[8] BORETTI, A. ; WATKINS, Simon: *Reduced Warm-Up and Recovery of the Exhaust and Coolant Heat with a Single Loop Turbo Steamer Integrated with the Engine Architecture in a Hybrid Electric Vehicle.* 2013-01-2827. SAE, 2013

[9] BREDEL, Eberhard ; NICKL, Jörg ; BARTOSCH, Stephan: Abwärmenutzung im Antrieb von Heute und Morgen. In: *MTZ* (2011), Nr. 04, S. 308–313

[10] BUNDESVERBAND DER DEUTSCHEN LUFTVERKEHRSWIRTSCHAFT ; BDL - BUNDESVERBAND DER DEUTSCHEN LUFTVERKEHRSWIRTSCHAFT E.V. (Hrsg.): *Klimaschutzreport 2016.* 2016

[11] CHEN, Huijuan ; GOSWAMI, Yogi ; RAHMAN, Muhammad ; STEFANAKOS, Elias: *Optimizing energy conversion using organic Rankine cycles and supercritical Rankine cycles.* ASME 5th International Conference on Energy Sustainability, 2011

[12] CHIEW, Lee ; CLEGG, Michael: *Waste Heat Energy Harvesting for Improving Vehicle Efficiency.* 2011-01-1167. SAE, 2011

[13] CLEMENTE, Stefano ; MICHELI, Diego ; REINI, Mauro ; TACCANI, Rodolfo: *Energy efficiency analysis of Organic Rankine Cycles with scroll expanders for cogenerative applications.* Bd. 97. 2012

[14] DAWIDZIAK, Johannes: *Methodische Entwicklung eines Systems zur Abgasenergierückgewinnung und dessen Untersuchung an einem Höchstleistungs-Dieselmotor.* Wiesbaden : Springer Vieweg, 2016 (Research). – ISBN 9783658110550

[15] DELLA BONA, Sebastian: *Aufbau und Automatisierung eines Prüfstandes zur Abgaswärmerückgewinnung im PKW*. Diplomarbeit. KIT, 2015

[16] DIESEL, Rudolf: *Theorie und Konstruktion eines rationellen Wärmemotors zum Ersatz der Dampfmaschine und der heute bekannten Verbrennungsmotoren*. Berlin : Springer, 1893

[17] DIETZEL, Fritz: *Turbinen, Pumpen und Verdichter*. Würzburg : Vogel-Verlag, 1980. – ISBN 3802301307

[18] DONN, Christian: *Untersuchung des thermischen Verhaltens und der Reibung an einem Diselmotor mit getrennter Kopf/Block-Kühlung*. Dissertation. Berlin : Logos, 2013

[19] EBNER, Lena: *Regulatory and Vehicle Related Environmental Conditions for Thermal Recuperation Systems in the Future*. Stuttgart : VDI Wissensforum, 2017 (3. VDI Fachkonferenz, Thermische Rekuperation in Fahrzeugen)

[20] EFFENBERGER, Helmut: *Dampferzeugung*. Berlin : Springer, 2000 (VDI-Buch). – ISBN 3540641750

[21] EISER, Axel ; DOERR, Joachim ; JUNG, Michael ; ADAM, Stephan: Der neue 1,8-l-TFSI-Motor von Audi: Teil 1: Grundmotor und Thermomanagement. In: *MTZ* (2011), Nr. 06, S. 466–474

[22] ENDO, T. ; KAWAJIRI, S. ; KOJIMA, Y. ; TAKAHASHI, K. ; BABA, T. ; IBARAKI, S. ; TAKAHASHI, T. ; SHINOHARA, M.: *Study on Maximizing Exergy in Automotive Engines*. SAE, 2007

[23] ERLANDSSON, Olof ; SKARE, Thomas ; CONTET, Arnaud: *On Handling Waste Heat from Waste Heat Recovery Systems in Heavy-Duty Vehicles*. 2015-01-2792. SAE, 2015

[24] EUROPÄISCHE KOMISSION: *VERORDNUNG (EU) 2016/427 DER KOMMISSION: zur Änderung der Verordnung (EG) Nr. 692/2008 hinsichtlich der Emissionen von leichten Personenkraftwagen und Nutzfahrzeugen (Euro 6)*. 2016

[25] EUROPÄISCHES PARLAMENT UND RAT: *Verordnung (EG) Nr. 443/2009*. 23. April 2009

[26] FARMER, G. T.: *Modern Climate Change Science: An Overview of Today's Climate Change Science*. 2015 (SpringerBriefs in Environmental Science). – ISBN 978-3-319-09221-8

[27] FRANKE, Andreas: *Thermische Rekuperation im instationären Betrieb – Ein Beitrag zur Optimierung des Clausius-Rankine-Prozesses zur Wärmerückgewinnung im Kraftfahrzeug*. Dissertation. Darmstadt : Technische Universität Darmstadt, 2016

[28] FREYMANN, Raymond ; RINGLER, Jürgen ; SEIFERT, Marco ; TILMANN, Horst: Der Turbosteamer der zweiten Generation. In: *MTZ* (2012), Nr. 02, S. 114–119

[29] FREYMANN, Raymond ; STROBL, Wolfgang ; OBIEGLO, Andreas: Der Turbosteamer: Ein System zur Kraft-Wärme-Kopplung im Automobil. In: *MTZ* (2008), Nr. 05, S. 404–412

[30] FRIEDRICH, Horst ; SCHIER, Michael ; HÄFELE, Christian ; WEILER, Tobias: Strom aus Abgasen - Fahrzeuggerechte Entwicklung thermoelektrischer Generatoren. In: *ATZ* (2010), Nr. 04, S. 292–299

[31] GAMMA TECHNOLOGIES, Inc.: *GT-SUITE: Software Documentation*. 2017

[32] GAMMA TECHNOLOGIES, Inc.: *GT - The Leading CAE Platform for Multi-Physics System Simulations: Waste Heat Recovery.* 2018. – URL https://www.gtisoft.com/ gt-suite-applications/thermal-management/waste-heat-recovery/. – Zugriffsdatum: 23.03.2018

[33] GÄRTNER, Jan ; KOCH, Thomas: *Kraftfahrzeug mit einer Abwärmenutzungsvorrichtung insbesondere zur Einspeisung der in nutzbare Leistung umgewandelten Abwärme in den Antrieb des Kraftfahrzeugs.* 2010. – URL https://google.com/patents/ DE102008060950A1?cl=de

[34] GERINGER, Bernhard ; HOFMANN, Peter ; HOLUB, Florian: Verbesserung des Hochlaufs und des Emissionsverhaltens im Kaltstart und Warmlauf bei Ottomotoren. In: *MTZ* (2010), Nr. 05, S. 368–372

[35] GETZLAFF, Uwe ; HENSEL, Sven ; REICHL, Sebastian: Simulation des Thermomanagements eines wassergekühlten Turboladers. In: *MTZ* (2010), Nr. 09, S. 598–601

[36] GHEBRU, Daniel: *Modellierung und Analyse des instationären thermischen Verhaltens von Verbrennungsmotor und Gesamtfahrzeug.* Bd. 2/2013. Dissertation. 2013. – ISBN 978-3-8325-3432-5

[37] GIERSCH, Hans-Ulrich: *Elektrische Maschinen.* 5., korrigierte Aufl. Stuttgart and Leipzig and Wiesbaden : Teubner, 2003 (Lehrbuch). – ISBN 3519468212

[38] GLENSVIG, M.: *Abwärmenutzung für Nutzfahrzeuge: Herausforderungen und Ergebnisse von Fahrzeugversuchen.* Stuttgart : 3. VDI Fachkonferenz: Thermische Rekuperation in Fahrzeugen, 2017

[39] GÜNTHER, Georg: *Comparison of exhaust gas turbo charging systems for truck applications in combination with an ORC for waste heat recovery.* Karlsruhe : Hectorschool KIT, 2017

[40] HAIDAR, Jihad ; GHOJEL, Jamil: *Waste Heat Recovery from the Exhaust of Low-Power Diesel Engine using Thermoelectric Generators.* 20th International Conference on Thermoelectrics, 2001

[41] HAN, Yongqiang ; JIANJIAN, Kang ; WANG, Xianfang ; CHEN, Yang ; HU, Zhichao: *The Effect of Cylinder Clearance on Output Work of ORC-FP used in Waste Energy Recovery.* 2014-01-2563. SAE, 2014

[42] HARTMANN, Andreas: *Energie- und Wärmemanagment mit thermischer Rekuperation für Personenkraftwagen.* Dissertation. Berlin : Logos, 2014

[43] HEBERLE, Florian: *Thermodynamik.* Bd. 22: *Untersuchungen zum Einsatz von zeotropen Fluidgemischen im Organic Rankine Cycle für die geothermische Stromerzeugung.* Berlin : Logos, 2013. – ISBN 9783832533557

[44] HEINLE, Dieter ; FEUERECKER, Günther ; STRAUSS, Thomas ; SCHMIDT, Michael: Zuheizsysteme: PTC-Zuheizer, Abgaswärmeübertrager, CO2-Wärmepumpen. In: *ATZ* (2003), Nr. 09, S. 846–851

[45] HEPKE, Georg: *Direkte Nutzung von Abgasenthalpie zur Effizienzsteigerung von Kraftfahrzeugen.* Dissertation. Technische Universität München, 2010

[46] HORST, Tilmann A.: *Betrieb eines Rankine-Prozesses zur Abgaswärmenutzung im PKW.* 1. Aufl. Herzogenrath : Shaker, 2015 (Berichte aus der Fahrzeugtechnik). – ISBN 978-3-8440-3923-8

[47] HORST, Tilmann A. ; TEGETHOFF, Wilhelm ; EILTS, Peter ; KOEHLER, Juergen: *Prediction of dynamic Rankine Cycle waste heat recovery performance and fuel saving potential in passenger car applications considering interactions with vehicles' energy management.* Bd. 78. 2014

[48] HUSCHER, Frederick: *Organic Rankine cycle turbine expander design, development, and 48 V mild hybrid system integration.* Baden Baden : ATZ 4th International Engine Congress, 2017

[49] JACOBS, Timothy: *Waste Heat Recovery Potential of Advanced Internal Combustion Engine Technologies.* ASME Journal of Energy Resources Technology, 2015

[50] KADUNIC, Samir: *Einfluss der Ladelufttemperatur auf den Ottomotor: Ein Potenzial zur Steigerung von Wirkungsgrad und Leistung aufgeladener Motoren.* 1. Aufl. 2015. Wiesbaden : Springer Fachmedien Wiesbaden GmbH, 2015. – ISBN 978-3-658-11135-9

[51] KADUNIC, Samir ; KIPKE, Peter ; WIEDEMANN, Bernd: *Potential of exhaust energy use for charge air cooling in supercharged diesel engines.* 2010-36-0478. SAE, 2010

[52] KADUNIC, Samir ; SCHERER, Florian ; BAAR, Roland ; ZEGENHAGEN, Tobias: Ladeluftkühlung mittels Abgasenergienutzung zur Wirkungsgradsteigerung von Ottomotoren. In: *MTZ* (2014), Nr. 01, S. 80–87

[53] KALIDE, Wolfgang: *Energieumwandlung in Kraft- und Arbeitsmaschinen: Kolbenmaschinen, Strömungsmaschinen, Kraftwerke.* 9. Aufl. München and Wien : Hanser, 2005. – ISBN 3446403965

[54] KIM, Young M. ; SHIN, Dong G. ; KIM, Chang G. ; CHO, Gyu B.: Single-loop organic Rankine cycles for engine waste heat recovery using both low- and high-temperature heat sources. In: *Elsevier Energy* (2016), Nr. 96, S. 482–494

[55] KLOSE, Arno ; KITTE, Jens ; BALS, Reinhold ; JÄNSCH, Daniel: *Wärmemanagement des Kraftfahrzeugs V: Potentialstudie verschiedener Wärmerekuperationskonzepte im Fahrzeug.* Tübingen : Expert Verlag, 2006

[56] KOCH, Thomas: *Numerischer Beitrag zur Charakterisierung und Vorausberechnung der Gemischbildung und Verbrennung in einem direkteingespritzten, strahlgeführten Ottomotor.* Dissertation. ETH Zürich, 2002

[57] KÖRNER, Jan E.: *Niedertemperatur-Abwärmenutzung mittels Organic-Rankine-Cycle im mobilen Einsatz.* Dissertation. Rostock : Audi-Dissertationsreihe, 2013

[58] KRALJEVIC, Ivica ; WEYHING, Thomas: *Rankine System mit Ruths-Speicher zu Abwärmerekuperation im Pkw.* Stuttgart : 3. VDI Fachkonferenz: Thermische Rekuperation in Fahrzeugen, 2017

[59] KRUSE, Alexander: *Thermoakustisch-elektrischer Generator (TAEG) zur Nutzung von Abgaswärme.* Nagold : Boysen Doktorandentage, 2016

[60] KUPFERSCHMID, Stefan: *Simulation unterschiedlicher Betriebsstrategien eines Abwärmerückgewinnungssystems und Modellierung der Auswirkungen auf den Betrieb eines schweren Nutzfahrzeugmotors.* Diplomarbeit. Karlsruhe : KIT, 2013

[61] LAGALY, Paul: *Vermessung eines Restwärmenutzungskreislaufs auf Basis eines Clausius-Rankine-Prozesses und Untersuchung der Wechselwirkung von Motor und Kreislauf.* Masterarbeit. Karlsruhe : KIT, 2016

[62] LANDESANSTALT FÜR UMWELT, MESSUNGEN UND NATURSCHUTZ BADEN-WÜRTTEMBERG: *Leitfaden Gewässerbezogene Anforderungen an Abwassereinleitungen.* Karlsruhe, 2015. – ISBN 978-3-88251-387-5

[63] LATZ, Gunnar: *Waste heat recovery from combustion engines based on the Rankine cycle.* Gothenburg, Sweden : Chalmers University of Technology, 2016

[64] LEE, B. ; JUNG, D. ; MYERS, J. ; KANG, J.-H. ; JUNG, Y.-H. ; KIM, K.-Y.: *Fuel Economy Improvement During Cold Start Using Recycled Exhaust Heat and Electrical Energy for Engine Oil and ATF Warm-Up.* 2014-01-0674. SAE, 2014

[65] LEIPERTZ, Alfred: *Technische Thermodynamik für Maschinenbauer, Fertigungstechniker, Verfahrenstechniker und Chemie- und Bioingenieure: = Engineering thermodynamics for mechanical engineers, production engineers, and chemical and bioengineers.* Als Ms. gedr., 3., korrigierte Aufl. Erlangen : ESYTEC, Energie- und Systemtechnik GmbH, 2006. – ISBN 3931901459

[66] LEMMON, E. ; HUBER, M.: *Reference Fluid Thermodynamic and Transport Properties Database (REFPROP).* Gaithersburg : National Institute of Standards and Technology (NIST), 2016

[67] LEMORT, Vincent ; QUOILIN, Sylvain ; CUEVAS, Cristian ; LEBRUN, Jean: Testing and modeling a scroll expander integrated into an Organic Rankine Cycle. In: *Elsevier Applied Thermal Engineering* 29 (2009), Nr. 14-15, S. 3094–3102

[68] LIEBL, Johannes ; NEUGEBAUER, Stephan ; EDER, Andreas ; LINDE, Mattias ; MAZAR, Boris ; STÜTZ, Wolfgang: Der thermoelektrische Generator von BMW macht Abwärme nutzbar. In: *MTZ* (2009), Nr. 04, S. 272–281

[69] LIU, Xiaobing ; YEBI, Adamu: *Real-Time embedded Implementation of Model Predictive Control of an Organic Rankine Cycle System.* SAE Thermalmanagement Systems Symposium, 2016

[70] LUCAS, Klaus: *Thermodynamik: Die Grundgesetze der Energie- und Stoffumwandlungen.* 6., bearb. Aufl. Berlin : Springer, 2007 (Springer-Lehrbuch). – ISBN 9783540735151

[71] MATOUSEK, Thomas: *Experimentelle Untersuchung und Energieflussanalyse verschiedener Thermomanagementmaßnahmen am Beispiel eines Audi Q7.* Diplomarbeit. KIT, 2012

[72] MATOUSEK, Thomas: *Kombinierte Nutzung von Abgaswärmestrom und Turbolader-Abwärme mittels eines Restwärmenutzungssystems nach Rankine an einem Ottomotor.* Stuttgart : 3. VDI Fachkonferenz: Thermische Rekuperation in Fahrzeugen, 2017

[73] MATOUSEK, Thomas ; BENS, Michael ; LAGALY, Paul ; KOCH, Thomas: Experimental Investigation of the Influence of Engine Operating Parameters on a Rankine Based Waste Heat Recovery System in a SI Engine. In: *SAE International Journal of Engines* (2018), Nr. 11, S. 147–160

[74] MATOUSEK, Thomas ; DAGEFÖRDE, Helge ; BERTSCH, Markus: *Influence of Injection Pressures up to 300 bar on Particle Emissions in a GDI-Engine.* Zürich : 17th International ETH-Conference on Combustion Generated Nanoparticles, 2013

[75] MATOUSEK, Thomas ; STAHL, Frank ; KOCH, Thomas: *Influence of engine operating parameters on the efficiency of a Rankine based waste heat recovery system and its behavior during cold start in a gasoline engine for automotive use.* Phoenix : SAE Thermalmanagement Systems Symposium, 2016

[76] MATOUSEK, Thomas ; STAHL, Frank ; KOCH, Thomas ; BENS, Michael: *Experimental investigation of the cold start behavior of different coolant circuits for a waste heat recovery system and their influence on the engine.* Baden Baden : ATZ 4th International Engine Congress, 2017

[77] MÜLLER, Rolf ; OECHSLEN, Holger ; SCHMIDT, Thomas ; DINGELSTADT, René ; EWERT, Sebastian: *Efficiency increase of the CV engine with a WHR turbine expander combined with a 48 V board net.* Baden Baden : ATZ 4th International Engine Congress, 2017

[78] NEUKIRCHNER, Heiko ; TORSTEN, Semper ; LÜDERITZ, Daniel ; DINGEL, Oliver: Symbiose aus Energierückgewinnung und Downsizing. In: *MTZ* (2014), Nr. 09, S. 14–21

[79] NEUNTEUFL, Klemens ; STEVENSON, Philip ; HÜLSER, Holger ; THEISSL, Helmut: Abwärmenutzung steigert Kraftstoffeffizienz. In: *MTZ* (2012), Nr. 12, S. 944–950

[80] PETERSON, R. B. ; WANG, H. ; HERRON, T.: Performance of a small-scale regenerative Rankine power cycle employing a scroll expander. In: *Proceedings of the Institution of Mechanical Engineers, Part A: Journal of Power and Energy* 222 (2008), Nr. 3, S. 271–282. – ISSN 0957-6509

[81] PLOETZ, Christiane ; REUSCHER, Günter ; ZWECK, Axel: *Mehr Wissen - weniger Ressourcen.* Düsseldorf : VDI Technologiezentrum GmbH, 2009

[82] PREISSINGER, Markus ; SCHWÖBEL, Johannes: *Design, Herstellung und Test eines idealen Rankine-Fluids für die Abgaswärmenutzung in der mobilen Anwendung.* Bericht zur Informationstagung Motoren. Frankfurt am Main : FVV, 2015

[83] RAMSBERGER, Florian ; ZEGENHAGEN, Tobias ; KADUNIC, Samir: Steigerung des Wirkungsgrads von Ottomotoren durch eine abgaswärmegetriebene Kälteanlage. In: *MTZ* (2010), Nr. 02, S. 122–127

[84] RANKINE, William John M.: Examples of the Application of the Second Law of Thermodynamics to a Perfect Steam-Engine and a Perfect Air-Engine. In: *Miscellaneous Scientific Papers* (1867), S. 439–453

[85] RISSE, Silvio ; ZELLBECK, Hans: Motornahe Abgasenergierekuperation bei einem Ottomotor. In: *MTZ* (2013), Nr. 01, S. 78–85

[86] RÖDER, Andreas: *Konstruktion, Aufbau und Validierung einer Anlage zur Nachbildung der Luftumströmung eines Pkw- Verbrennungsmotors und dessen Abgasanlage auf einem Motorprüfstand.* Bachelorarbeit. KIT, 2016

[87] ROEDDER, Maximilian ; NEEF, Matthias ; LAUX, Christoph ; PRIEBE, Klaus-P.: Systematic Fluid Selection for Organic Rankine Cycles and Performance Analysis for a Combined High and Low Temperature Cycle. In: *Journal of Engineering for Gas Turbines and Power* 138 (2016), Nr. 3, S. 8–17. – ISSN 0742-4795

[88] SAHOO, Dipankar ; KOTRBA, Adam ; STEINER, Tom ; KLOPFER, Ron ; SWIFT, Greg: *Waste Heat Recovery for Class-8 Heavy Duty Truck Using Thermoacoustic Converter (TAC) Technology.* Phoenix : SAE Thermalmanagement Systems Symposium, 2016

[89] SCHNEIDER, Tobias: *Aufbau eines Motorenprüfstands zur Untersuchung von Kreisprozessen zur Restwärmenutzung für den Einsatz in zukünftigen Kraftfahrzeugen.* Bachelorarbeit. KIT, 2014

[90] SCHOBLICK, Robert: *Antriebe von Elektroautos in der Praxis: Motoren · Batterietechnik · Leistungstechnik.* Haar : Franzis Verlag, 2013 (Elektronik). – ISBN 9783645270311

[91] SCHUMANN, Florian: *Experimentelle Grundlagenuntersuchungen zum Katalysatorheizbetrieb mit strahlgeführter Benzindirekteinspritzung und Einspritzdrücken bis 800 bar.* Dissertation. Berlin : Logos, 2014

[92] SEIFERT, Marco ; RINGLER, Jürgen ; VIANNEY, Guyotot ; RAYMOND, Freymann: *Potenzial der Abwärmerückgewinnung mittels eines Rankine-Prozesses beim PKW.* Graz : 12. Tagung Der Arbeitsprozess des Verbrennungsmotors, 2009

[93] SHUDO, T. ; TOSHINAGA, K.: Combustion control for waste-heat recovery system in internal combustion engine vehicles: Increase in exhaust-gas heat by combustion phasing and its effect on thermal efficiency factors. In: *International Journal of Engine Research* (2010), Nr. 2, S. 99–108. – ISSN 1468-0874

[94] SONG, Binyang ; ZHUGE, Weilin ; YIN, Yong: *Parameter Study of a Brayton Cycle Waste Heat Recovery System for Turbocharged Diesel Engines.* Incline Village, Nevada, USA : ASME 2013 Fluids Engineering Division Summer Meeting, 2013

[95] SOURELL, R. ; KUNTE, H. ; HÜTKER, J. ; NIKOLOV, A.: *Expansionsmaschine (Vorstudie).* Bd. Heft 1044. Abschlussbericht. Frankfurt am Main : FVV, 2014

[96] SPICHER, Ulrich ; MATOUSEK, Thomas: *Energiebedarf und CO2-Emissionen von konventionellen und neuen Kraftfahrzeugantrieben unter Alltagsbedingungen.* Fachtagung. Wolfsburg : ATZ live: Der Antrieb von morgen, 2014

[97] SPIEGEL PRINT REDAKTION: Automobile: Wie praxistauglich und umweltfreundlich ist ein schneller Umstieg auf Elektroautos wirklich? In: *Der Spiegel* (2017), Nr. 34, S. 120–121

[98] STAHL, Frank: *Development of a feed pump for closed Rankine-circuits in mobile applications with multiple heat sources.* Braunschweig : WKM Symposium, 2017

[99] STAHL, Frank: *Systematische Entwicklung von Förderpumpen für mobile Rankinekreisläufe.* Stuttgart : 3. VDI Fachkonferenz: Thermische Rekuperation in Fahrzeugen, 2017

[100] STEPHAN, Peter ; SCHABER, Karlheinz ; STEPHAN, Karl ; MAYINGER, Franz: *Thermodynamik: Grundlagen und technische Anwendungen Band 1: Einstoffsysteme.* 19., ergänzte Aufl. 2013. Berlin : Springer Berlin, 2013 (Springer-Lehrbuch). – ISBN 978-3-642-30097-4

[101] STRAUSS, Karl: *Kraftwerkstechnik: zur Nutzung fossiler, nuklearer und regenerativer Energiequellen.* 7. Berlin : Springer Vieweg, 2016

[102] STRUZYNA, Ralf: *Kreisprozesse: Nutzung der Motorabwärme durch Kreisprozesse.* Bd. Heft 997. Abschlussbericht. Frankfurt am Main : FVV, 2013

[103] STRUZYNA, Ralf ; MENNE, Andreas: *Definition und Erprobung von Fluiden zum Einsatz in Waste-Energy-Recovery-Anlagen.* Magdeburg : FVV-Frühjahrtagung, 2014

[104] TAUVERON, Nicolas ; COLASSON, Stephane ; GRUSS, Jane-Antoine: *Available Systems for the Conversion of Waste Heat to Electricity.* Proceedings. Montreal, Quebec, Cancada : ASME 2014 International Mechanical Engineering Congress and Exposition, 2014

[105] THOMA, Werner ; GÄRTNER, Jan ; KÖHLER, Jürgen: *Der Ejektorkreislauf zur alternativen Ladeluftkühlung.* Haus der Technik, 2016 (Wärmemanagement des Kraftfahrzeugs X: Energiemanagement)

[106] UMWELT BUNDESAMT ; UMWELT BUNDESAMT (Hrsg.): *Indikator: Umweltfreundlicher Personenverkehr.* 2017. – URL http://www.umweltbundesamt.de/indikator-umweltfreundlicher-personenverkehr#textpart-1. – Zugriffsdatum: 25.03.2019

[107] VELJI, Amin ; YEOM, K. ; WAGNER, Uwe ; ROSSBACH, M. ; SUNTZ, Rainer ; HENNING, Bockhorn: *Investigations of the formation and oxidation of soot inside a direct injection spark ignition engine using advanced Laser-Techniques.* 2010-01-0352. Detroid, Mi,, USA : SAE 2010 World Congress & Exhibition, 2010

[108] VERBAND DER AUTOMOBILINDUSTRIE E. V.: *Standardaufheizung für Pkw mit 1 bis 2 Sitzreihen: VDA 220.* VDA-Empfehlung. Verband der Automobilindustrie e. V. (VDA), 2005

[109] WANG, Jianyong ; WANG, Jianfeng ; PAN, Zhao ; DAI, Yiping ; PENG, Yan: *Thermodynamic Analysis and Comparison Study of an Organic Rankine Cycle (ORC) and a Kalina Cycle for Waste Heat Recovery of Compressor Intercooling.* ASME Turbo Expo 2014: Turbine Technical Conference and Exposition, 2014

[110] WIEHN, H ; TRASSL, E. H. W. ; SCHULENBERG, F. J. ; SCHWARZBACH, K.: *Strom aus Steinkohle: Stand der Kraftwerkstechnik.* Berlin : Springer. – ISBN 978-3-642-50251-4

[111] ZELLBECK, Hans ; RISSE, Silvio ; TILO, Roß: *Thermische Rekuperation im Kontext zur Avgasturboaufladung und Abgasnachbehandlung.* Tübingen : Expert Verlag, 2009

[112] ZIMMERLE, Daniel ; CIRINCIONE, Nicholas: *Analysis of Performance of Direct Dry Cooling for Organic Rankine Cycle Systems.* Washington, DC, USA : ASME 5th International Conference on Energy Sustainability, 2011

Lebenslauf

Persönliche Daten

Name:	Thomas Matoušek
Geburtsdatum:	27.08.1986
Geburtsort:	Erlangen

Schulbildung

1993 - 1997	Grundschule Heroldsberg
1997 - 2006	Gymnasium Eckental

Studium

2006 - 2008	Maschinenbaustudium an der Universität Erlangen
2008 - 2012	Maschinenbaustudium an der Universität Karlsruhe (TH)

Berufliche Laufbahn

2012 - 2014	Wissenschaftlicher Mitarbeiter der MOT GmbH
2014 - 2017	Wissenschaftlicher Mitarbeiter des Institut für Kolbenmaschinen, Karlsruher Institut für Technologie
seit 2017	Entwicklungsingenieur der Audi AG